Forschungsberichte aus dem Lehrstuhl für Regelungssysteme
Technische Universität Kaiserslautern

Band 15

Forschungsberichte aus dem Lehrstuhl für Regelungssysteme

Technische Universität Kaiserslautern

Band 15

Herausgeber:

Prof. Dr. Steven Liu

Felix Berkel

Contributions to Event-triggered and Distributed Model Predictive Control

Logos Verlag Berlin

λογος

Forschungsberichte aus dem Lehrstuhl für Regelungssysteme
Technische Universität Kaiserslautern

Herausgegeben von
Univ.-Prof. Dr.-Ing. Steven Liu
Lehrstuhl für Regelungssysteme
Technische Universität Kaiserslautern
Erwin-Schrödinger-Str. 12/332
D-67663 Kaiserslautern
E-Mail: sliu@eit.uni-kl.de

Bibliographic information published by the Deutsche Nationalbibliothek

The Deutsche Nationalbibliothek lists this publication in the Deutsche Nationalbibliografie; detailed bibliographic data are available on the Internet at http://dnb.d-nb.de .

ISBN 978-3-8325-4935-0
ISSN 2190-7897

Logos Verlag Berlin GmbH
Comeniushof, Gubener Str. 47,
10243 Berlin
Tel.: +49 (0)30 / 42 85 10 90
Fax: +49 (0)30 / 42 85 10 92
http://www.logos-verlag.de

Contributions to Event-triggered and Distributed Model Predictive Control

Beiträge zu ereignisbasierter und verteilter modellprädiktiver Regelung

Vom Fachbereich Elektrotechnik und Informationstechnik
der Technischen Universität Kaiserslautern
zur Verleihung des akademischen Grades
Doktor der Ingenieurwissenschaften (Dr.-Ing.)
genehmigte Dissertation

von

Dipl.-Ing. Felix Berkel
geboren in Ludwigshafen am Rhein

D 386

Tag der mündlichen Prüfung:	14.02.2019
Dekan des Fachbereichs:	Prof. Dr.-Ing. Ralph Urbansky
Vorsitzender der Prüfungskommision:	Prof. Dr.-Ing. Norbert Wehn
1. Berichterstatter:	Prof. Dr.-Ing. Steven Liu
2. Berichterstatter:	Prof. Dr.-Ing. Olaf Stursberg

Acknowledgment

Foremost, I would like to thank Prof. Dr.-Ing. Steven Liu for his support during writing this thesis. He has always given me the freedom to work on problems of my choosing and encouraged the publication of my results. I also appreciated the possibilities to attend conferences and workshops as well as to visit other universities to present my results and learn from other researchers. I would like to thank Prof. Dr.-Ing. Olaf Stursberg for his interest in my thesis and joining the thesis committee as a reviewer. Thanks also go to Prof. Dr.-Ing. Norbert Wehn for joining the thesis committee as the chair.

Special thanks go to my former colleagues Markus Bell, Sebastian Caba, Fabian Kennel and Tim Steiner in the "upstairs office", where, despite the fun we had, I really benefited from the many discussions and collaborations. I would also like to thank Alen Turnwald who made the time commuting to Kaiserslautern much more pleasant. Moreover, I would like to thank Markus Lepper and Jawad Ismail for the more than welcome distractions through the many discussions about sports.

I would like to thank my former colleagues Sanad Al-Areqi, Xiang Chen, Pedro Dos Santos, Filipe Figueiredo, Daniel Görges, Yanhao He, Muhammad Ikhsan, Hafiz Kashif Iqbal, Peter Müller, Tim Nagel, Sven Reimann, Stefan Simon, Christian Tuttas, Yun Wan, Hengyi Wang, Jianfei Wang, Benjamin Watkins, Min Wu, Wei Wu and Yakun Zhou. Working with them has always been a great pleasure. Thanks also go to Swen Becker and Thomas Janz as well as Jutta Lenhardt for providing a very good technical and administrative environment.

I would like to express my gratitude to my parents and family who have been very supportive over the past years. Finally, I would like to thank my wife Kathrin for her encouragement and endless patience, especially in the last months finishing this thesis. I really admire your positive attitude, stamina and eye for the important things in life.

Contents

Notation

Sets

$\mathbb{N}$	set of non-negative integers,
$\mathbb{Z}$	set of integers,
$\mathbb{R}$	set of real numbers,
$\mathbb{C}$	set of complex numbers,
$\mathbb{R}^n$	set of real vectors of length n,
$\mathbb{R}^{m\times n}$	set of real matrices of dimension $m \times n$,
$\mathbb{Z}_{n_1:n_2}$	set of integers from n_1 to n_2,
$\emptyset$	empty set.

Algebraic Operators

Let $z_1, z_2 \in \mathbb{Z}$, $z_2 \neq 0$ as well as $s \in \mathbb{R}$ be scalars, $\boldsymbol{v}, \boldsymbol{x} \in \mathbb{R}^n$ be vectors, $\boldsymbol{M} \in \mathbb{R}^{n\times n}$ be a symmetric matrix and $\mathcal{S} \subseteq \mathbb{R}^n$ a set.

$[z_1]_{z_2}$	modulo operation,
$\lvert s\rvert$	absolute value,
$\lVert\boldsymbol{v}\rVert$	Euclidean vector norm,
$\lVert\boldsymbol{v}\rVert_\infty$	infinity vector norm,
$\lVert\boldsymbol{v}\rVert_\mathcal{S}$	the distance $\lVert\boldsymbol{v}\rVert_\mathcal{S} = \min_{\boldsymbol{x}\in\mathcal{S}} \lVert\boldsymbol{v}-\boldsymbol{x}\rVert$,
$\boldsymbol{M} \succ \boldsymbol{0}$	matrix is positive definite,
$\boldsymbol{M} \succeq \boldsymbol{0}$	matrix is positive semidefinite,
$\lambda^{\mathrm{m}}(\boldsymbol{M})$	minimum eigenvalue of a positive semidefinite matrix $\boldsymbol{M}$,
$\lambda^{\mathrm{M}}(\boldsymbol{M})$	maximum eigenvalue of a positive semidefinite matrix $\boldsymbol{M}$.

Let $\underline{v} \in \mathbb{C}$ be a complex scalar.

$\Re\{\underline{v}\}$	real part,
$\Im\{\underline{v}\}$	imaginary part,
$\lvert\underline{v}\rvert$	absolute value,
$\overline{\underline{v}}$	conjugate complex value.

Set Operators

Let $\mathcal{N} \subset \mathbb{Z}$ as well as $\mathcal{S}_1, \ldots, \mathcal{S}_M \subseteq \mathbb{R}^n$ be sets and $\boldsymbol{M} \in \mathbb{R}^{m\times n}$ as well as $\boldsymbol{N} \in \mathbb{R}^{n\times m}$ be matrices.

$\mathcal{N}(i)$	ith element of $\mathcal{N}$,
$\lvert\mathcal{S}_1\rvert$	cardinality of $\mathcal{S}_1$,
$\mathcal{S}_1 \oplus \mathcal{S}_2$	Minkowski sum of sets, $\mathcal{S}_1 \oplus \mathcal{S}_2 = \{\boldsymbol{s}_1 + \boldsymbol{s}_2 \mid \boldsymbol{s}_1 \in \mathcal{S}_1, \boldsymbol{s}_2 \in \mathcal{S}_2\}$,
$\mathcal{S}_1 \ominus \mathcal{S}_2$	Pontryagin difference of sets, $\mathcal{S}_1 \ominus \mathcal{S}_2 = \{\boldsymbol{s}_1 \mid \{\boldsymbol{s}_1\} \oplus \mathcal{S}_2 \subseteq \mathcal{S}_1\}$,
$\mathcal{S}_1 \cap \mathcal{S}_2$	intersection of sets, $\mathcal{S}_1 \cap \mathcal{S}_2 = \{\boldsymbol{s} \mid \boldsymbol{s} \in \mathcal{S}_1 \text{ and } \boldsymbol{s} \in \mathcal{S}_2\}$,
$\mathcal{S}_1 \cup \mathcal{S}_2$	union of sets, $\mathcal{S}_1 \cup \mathcal{S}_2 = \{\boldsymbol{s} \mid \boldsymbol{s} \in \mathcal{S}_1 \text{ or } \boldsymbol{s} \in \mathcal{S}_2\}$,
$\mathcal{S}_1 \setminus \mathcal{S}_2$	set difference of sets, $\mathcal{S}_1 \setminus \mathcal{S}_2 = \{\boldsymbol{s} \mid \boldsymbol{s} \in \mathcal{S}_1 \text{ and } \boldsymbol{s} \notin \mathcal{S}_2\}$,
$\mathcal{S}_1 \times \mathcal{S}_2$	cartesian product of sets, $\mathcal{S}_1 \times \mathcal{S}_2 = \left\{ \left[\boldsymbol{s}_1^{\mathrm{T}}, \boldsymbol{s}_2^{\mathrm{T}}\right]^{\mathrm{T}} \mid \boldsymbol{s}_1 \in \mathcal{S}_1, \boldsymbol{s}_2 \in \mathcal{S}_2 \right\}$,
$\bigoplus_{i=1}^{M} \mathcal{S}_i$	Minkowski sum of M sets, $\bigoplus_{i=1}^{M} \mathcal{S}_i = \mathcal{S}_1 \oplus \mathcal{S}_2 \oplus \ldots \oplus \mathcal{S}_M$,
$\prod_{i=1}^{M} \mathcal{S}_i$	cartesian product of M sets, $\prod_{i=1}^{M} \mathcal{S}_i = \mathcal{S}_1 \times \mathcal{S}_2 \times \ldots \times \mathcal{S}_M$,
$\bigcup_{i=1}^{M} \mathcal{S}_i$	union of M sets, $\bigcup_{i=1}^{M} \mathcal{S}_i = \{\boldsymbol{s} \mid \boldsymbol{s} \in \mathcal{S}_1 \text{ or ... or } \boldsymbol{s} \in \mathcal{S}_M\}$,
$\bigcap_{i=1}^{M} \mathcal{S}_i$	intersection of M sets, $\bigcap_{i=1}^{M} \mathcal{S}_i = \{\boldsymbol{s} \mid \boldsymbol{s} \in \mathcal{S}_1 \text{ and ... and } \boldsymbol{s} \in \mathcal{S}_M\}$,
$\boldsymbol{M}\mathcal{S}_1$	linear map, $\boldsymbol{M}\mathcal{S}_1 = \{\boldsymbol{M}\boldsymbol{s} \mid \boldsymbol{s} \in \mathcal{S}_1\}$,
$\boldsymbol{N}^{-1}\mathcal{S}_1$	inverse linear map, $\boldsymbol{N}^{-1}\mathcal{S}_1 = \{\boldsymbol{s} \mid \boldsymbol{N}\boldsymbol{s} \in \mathcal{S}_1\}$.

Others

$\mathbf{0}$	zero matrix of appropriate dimension,
$\boldsymbol{I}$	identity matrix of appropriate dimension,
$\mathbf{1}$	vector of appropriate dimension with all entries equal to one,
$\left(\begin{smallmatrix} \boldsymbol{A} & \boldsymbol{B} \\ * & \boldsymbol{C} \end{smallmatrix}\right)$	symmetric matrix $\left(\begin{smallmatrix} \boldsymbol{A} & \boldsymbol{B} \\ \boldsymbol{B}^{\mathrm{T}} & \boldsymbol{C} \end{smallmatrix}\right)$,
$\boldsymbol{z}^*$	variable $\boldsymbol{z}$ determined by optimization.

Let $\boldsymbol{M}_{ij} \in \mathbb{R}^{n_i \times m_j}$ be matrices.

$(\boldsymbol{M}_{ij})$	matrix consisting of blocks $\boldsymbol{M}_{ij}$.

Let $\boldsymbol{v}_i \in \mathbb{R}^{n_i}$ and $\boldsymbol{M}_i \in \mathbb{R}^{n_i \times m_i}$ with $i \in \mathbb{Z}_{1:M}$ be vectors and matrices, respectively.

$\mathrm{diag}_{i\in\mathbb{Z}_{1:M}}(\boldsymbol{M}_i)$	block-diagonal matrix with blocks $\boldsymbol{M}_i$,
$\mathrm{col}_{i\in\mathbb{Z}_{1:M}}(\boldsymbol{v}_i)$	vector obtained from stacking the vectors $\boldsymbol{v}_i$.

Acronyms

A2A	All-to-All
ADMM	Alternating Direction Method of Multipliers
CMPC	Centralized Model Predictive Control
C2A	Controller-to-Actuator
DeMPC	Decentralized Model Predictive Control
DMPC	Distributed Model Predictive Control
DOA	Distributed Optimization Algorithm
FBS	Forward Backward Sweep
EMPC	Economic Model Predictive Control
GAS	Globally Asymptotically Stable
GES	Globally Exponentially Stable
LMI	Linear Matrix Inequality
LP	Linear Program
LPTV	Linear Periodically Time-Varying
LTI	Linear Time-Invariant
LV	Low Voltage
MIQP	Mixed-Integer Quadratic Program
MPC	Model Predictive Control
MV	Medium Voltage
NCS	Networked Control System
N2N	Neighbor-to-Neighbor
OB-MPC	Output-Based Model Predictive Control
PI	Positive Invariant
PPI	Periodically Positive Invariant
PV	Photovoltaic
QP	Quadratic Program
QPQC	Quadratically Constrained Quadratic Program
RPI	Robustly Positive Invariant
ROA	Region Of Attraction
S2C	Sensor-to-Controller
WDN	Water Distribution Network
ZOH	Zero-Order Hold

1.1 Event-triggered and Networked MPC

1.1.1 Motivation

A Networked Control System (NCS) consists of usually spatially distributed sensors, actuators and controllers that communicate over a shared communication network. An example for a single-loop NCS set-up is given in Figure 1.1. This configuration provides many advantages, such as reduced installation and maintenance costs as well as increased flexibility and reconfigurability, compared to conventional hard-wired control systems. The advantages of NCSs have led to the fact that they can be found in various applications, such as in automotive, marine and flight control [JTN05] or manufacturing and process control [MT07]. General overviews of NCSs can be found, e.g., in [HNX07, Zam08, GC10].

On the contrary, a spare usage of the network is desired since energy and communication resources in the network and its nodes are usually limited. Thus, it is reasonable to consider the network utilization in the controller design to provide a compromise between control performance and communication effort. One methodology which enables this in a systematic way is event-triggered control. Here the information transmission is not initiated in a time-triggered fashion, i.e., at predefined time instants, but by an event trigger. It decides when an update of information is required from a performance or stability perspective. Technically, this means that events are triggered when the deviation of the plant's behavior to a previously communicated or predicted behavior violates some kind of threshold. More detailed introductions and recent surveys on event-triggered control can be found in [HJT12, Cas14, MM15].

Another aspect to be taken into account in NCSs is that the network induces different types of imperfections into the control loop. Those imperfections are, for instance, time-

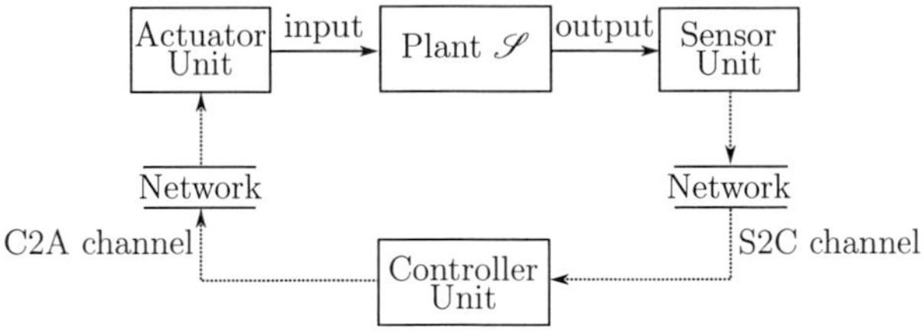

Figure 1.1: The figure shows a single control loop closed over a network. The sensor unit transmits the measured output over the S2C channel of the network. The controller unit sends the input over the C2A channel to the actuator.

varying transmission delays and sampling instants as well as medium access constraints, packet dropouts or quantization errors. They need to be taken into account in the controller design so that the stability of the closed-loop system is not deteriorated. In the survey papers [HNX07, Zam08] good overviews of network-induced imperfections are provided.

1.1.2 Related Work

In recent years, the event-triggering idea has also been applied to MPC. Moreover, networked MPC, i.e., MPC considering network-induced effects, has been investigated in the literature. In the sequel event-triggered MPC approaches are discussed first. Then, networked MPC schemes are reviewed. The discussion in this section is restricted to centralized single-loop set-ups. An overview of approaches for distributed control set-ups is included in the next section.

In the majority of event-triggered MPC approaches, the event triggers monitor the deviation between actual and predicted state which is obtained using the nominal model, i.e., a(n) (undistorted) reference model. While in [BB12, LHJ13, LS14, BHA17] set-based thresholds are applied, in [VFK$^+$10, EDK11a, EDK11b, HS15] events are triggered whenever the decrease of the Lyapunov function cannot be guaranteed. In [HS15] also an additional relaxed condition exploiting a non-monotonic Lyapunov function approach is presented. The authors in [HAD15] develop a dual-mode approach which does not check the decrease of the cost function. A benefit is that the trigger does not rely on parameters which might be unknown in an application, such as the Lipschitz constant of the stage cost in an unconstrained set-up.

It is worth mentioning that some of the above-listed publications consider special cases of the set-up depicted in Figure 1.1. Although it is not mentioned explicitly there, the approaches (except from [BB12]) can be applied to the set-up shown in Figure 1.1 when the predicted nominal state sequence is sent to the sensor unit and the event-triggering condition is checked there. A different methodology can be found in [IFM17] where the actuator and sensor unit are located at the same place and run a nominal model (which is of course also known by the controller) of the plant inside.

In [JDM15] explicit MPC for linear systems with its piecewise-affine control law is employed and events are triggered when the state enters a new set of the polytopic partitioning of the state space. Moreover, a suboptimal control law is presented where a predefined decrease of the cost function is ensured.

The above-mentioned approaches consider the state-feedback case, i.e., assume that the complete state of the plant can be measured. However, in many practical applications only output information is available (output-based or output-feedback case). In [LZN17] an Output-Based Model Predictive Control (OB-MPC) approach for an event-triggered computation of the control law is presented. In [ZLL14] a state- and output-feedback event-triggered economic MPC is considered. Both approaches assume that the output measurement is available at the observer at every sampling instant. The approaches in [SLH10] and [KF16] consider set-ups where the observer is placed at the controller side and utilize event-triggered communication of the output measurement over the S2C channel.

Considering networked MPC for the control over networks with transmission delays and packet dropouts, MPC has shown to be a well-suited methodology due to the inherent availability of predicted state and input sequences which can be used when information is delayed or lost. The approaches in [PP11] and [VF11] make use of a buffered actuator and a network compensator in the controller node which employs a model of the plant to compensate delays and packet dropouts. In [NVKP14] a system with polytopic uncertainty is controlled over a network with packet dropouts in the S2C and C2A channel. All above-mentioned approaches are deterministic and make an assumption on the maximum number of packet dropouts in a row. In [QMFC15] a stochastic controller with random packet dropouts according to some stochastic process in the C2A channel is designed.

The output-based case for unconstrained linear quadratic Gaussian control with packet dropouts is considered in [SSF+07]. An OB-MPC with packet dropouts in the S2C-channel is proposed in [LS13b].

An overview of the discussed literature on MPC for NCSs is given in Table 1.1. For reasons of completeness the work of [EDK11b] is added at this point although it considers decentralized set-ups which are introduced in the next section.

Table 1.1: Overview of literature on networked and event-triggered MPC. The columns S2C and C2A ET-channel say if the communication over this channel is event-triggered in the considered publication.

Paper	ET-channel		Feedback	Network-induced effects	Control architecture
	S2C	C2A			
[VFK⁺10, EDK11a, EDK11b, LHJ13, LS14, HS15, HAD15, IFM17, BHA17]	✓	✓	state	-	centralized
[BB12]	✓	-	state	-	centralized
[EDK11a]	✓	✓	state	-	decentralized
[ZLL14, LZN17]	-	✓	output	-	centralized
[SLH10, KF16]	✓	-	output	-	centralized
[PP11, VF11]	-		state	✓	centralized
[LS13b]	-		output	✓	centralized

1.2 Distributed MPC

1.2.1 Motivation

Centralized control of systems with many spatially distributed sensors and actuators is often impossible or undesirable. The information of all sensors in the system has to be gathered at the central controller, the control law has to be computed and the input has to be sent back to the actuators. This job can be demanding from a communicational and computational point of view. Moreover, the information gathering might be undesired, for example, when the overall plant is owned by different operators which do not want to share specific information with each other. Systems which are controlled in a centralized fashion are also vulnerable to failures, in the sense that the overall system can no longer be controlled in the event of a malfunction in the centralized controller.

One possible remedy is the decomposition of the plant into several coupled subsystems which are individually controlled. The coupling between the subsystems can be of different nature. It can appear in the dynamics, i.e., the dynamics of a subsystem are influenced by the inputs or states of other subsystems. Alternatively, the coupling can occur in a common constraint that the subsystems have to satisfy. Another coupling source is a common cost function to be minimized by the controllers of the subsystems.

If the controllers of the subsystems do not share any information with each other, this for decomposition-based approaches is to solve the dual optimization problem in an event-triggered fashion in the sense that primal and dual optimization variables are only exchanged between the subsystems when it is required for the convergence of the optimization algorithm. In [RG17] a concept similar to [MUA14] is applied to DMPC.

The approaches mentioned up to now have considered MPC schemes where a set-point is stabilized. Usually these set-points are known a priori or computed by an upper level optimization scheme which minimizes an economic objective. This is a well-known concept, for instance, applied in process industry [Sca09] or for control of a Water Distribution Network (WDN) [OBPB12]. In Economic Model Predictive Control (EMPC) a single-layer controller structure is applied where the economic objective is incorporated into the controller. Technically, this means that the cost function of the MPC reflects an economical goal rather than a function which is positive definite to some reference value. An introduction to EMPC can be found in [RAB12], a recent survey is given in [MA17]. A distributed EMPC scheme is presented in [LA14]. Distributed EMPC with applications to WDNs is investigated in [GOMP17].

An overview of the discussed event-triggered DMPC approaches can be found in Table 1.2.

Table 1.2: Overview of literature on DMPC approaches with event-triggered communication

Paper	Plant model	Coupling	Communication style	Objective
[GS16, RG17]	LTI	cost, dynamics	iterative	regulation
[GS13b]	hybrid	cost, constraints	iterative	regulation
[LZL+13]	nonlinear	cost, dynamics	iterative	regulation
[EDK12, HAD14, LYSW15, ML18]	nonlinear	cost	iterative	regulation

1.3 Outline and Contributions

As outlined in the short literature discussion in Section 1.1, the approaches for event-triggered MPC of NCSs have so far mostly considered control systems with state feedback. In practice, however, it is usually the case that the full state is not measured, but only certain outputs. As pointed out in the survey paper [HJT12], output-based approaches are hard to design and optimize as the separation principle does not hold. In addition, the treatment of network-induced effects in output-based event-triggered methods has not yet been investigated. This work intends to contribute to closing these existing gaps.

When considering the literature review on DMPC in Section 1.2, it becomes obvious that many different DMPC approaches have been proposed but little attention has been paid to the communication effort between the subsystems so far. In particular, the idea of event-triggered communication for its reduction has only been applied for certain set-ups and algorithms. A further aim of this work is to close the existing gap by transferring the idea of event-triggered communication to other classes of algorithms.

In the following, the exact contents of the chapters of the thesis are presented. In addition, the contributions of the work, which are thematically assigned to the chapters, are detailed.

Chapter 2

This chapter focuses on the background material required for the thesis. Basic mathematical definitions and existing concepts related to system and control theory are revisited in Section 2.1 and 2.2, respectively. Moreover, Section 2.3 introduces required MPC-related concepts, including robust formulations for LTI systems and economic formulations for Linear Periodically Time-Varying (LPTV) systems.

Chapter 3

This chapter introduces the set-ups that are considered throughout the thesis. The architecture of decentralized control for NCSs and the set-up of distributed control of coupled subsystems are presented in Section 3.1. Section 3.2 deals with the considered plant models: systems consisting of coupled LTI and LPTV subsystems subject to constraints. The model of the communication network is presented in Section 3.3. In Section 3.4 a suboptimal robust MPC scheme is introduced which is required for the analysis of some of the event-triggered and distributed MPC schemes.

boundaries by joined coordination of the producers and consumers in the grid. A control-oriented model which reflects the major properties of the system and allows at the same time efficient calculations is derived. It consists of subsystems with decoupled LPTV dynamics subject to decoupled but non-convex constraints. Moreover, the subsystems are coupled in a common constraint. The considered scenario is interesting for practical applications since one controlled consumer is a WDN which belongs to the same operator as the power grid and thus synergies effects can be exploited.

A DMPC scheme suitable for both applications minimizing an economic objective and running the simplified models inside is designed. It is tailored to the non-convexity as well as to the special structure in coupling dynamics and constraints. From a practical point of view, the main motivation for a distributed implementation is that it provides increased robustness against failures compared to a centralized implementation. This is an important property for the considered infrastructure systems. To reduce the communication load, the algorithm uses event-triggered communication between the subsystems. The scheme is derived in a general form and can thus be applied to other systems having this structure.

Beside the contributions from the point of view of the applications, which are detailed in the chapter, the approach contributes from a methodological point of view to event-triggered DMPC of LPTV systems with economic cost function (cf. Table 1.2).

Chapter 8

Conclusions on the content of the thesis are drawn in Section 8.1. Moreover, possible directions for future research are provided in Section 8.2.

Appendices

Some supplementary material which is repeatedly required for the proofs and information about the implementation is provided in Appendix A. For reasons of clarity, some proofs of mathematical statements have been moved to Appendix B.

Summary

Table 1.3 summarizes the considered plant model and controller set-up of the main chapters. The table is extended with deepened insights into the methodologies in Chapter 8 (see Table 8.1).

Table 1.3: Overview of considered plant model and controller set-up

Chapter	4	5	6	7
Plant model	discrete-time & sampled-data LTI	discrete-time LTI	discrete-time LTI	discrete-time LPTV
Network-induced effects	✓	-	-	-
Dynamic coupling	state & inputs	state	state	input/-
Constraint coupling	-	-	-	✓
Constraints	convex	convex	convex	non-convex
Feedback	output	state	state	state
Control architecture	decentralized	distributed	distributed	distributed
Objective function	regulation	regulation	regulation	economic
Controller attitude	non-cooperative	non-cooperative	cooperative	cooperative
Communication style	-	non-iterative	iterative	iterative
Purpose event trigger	communication S2C and C2A	communication N2N	cooperation N2N	communication N2N/A2A

1.4 Publications

Several publications on model predictive, event-triggered and distributed control have been finished during the doctoral studies. A chronological list of the articles and their relation to the topics as well as to the thesis is given in the following.

[BGL13] F. Berkel, D. Görges, S. Liu: Load-Frequency Control, Economic Dispatch and Unit Commitment in Smart Mircogrids based on Hierarchical Model Predictive Control. In: *Proceedings of the 52nd IEEE Conference on Decision and Control*, 2013, p.2326-2333 (hierarchical MPC of power networks)

[BFB+16] M. Bell, S. Fuchs, F. Berkel, S. Liu, D. Görges: A Privacy Preserving Negotiation-Based Control Scheme for Low Voltage Grids. In: *Proceedings of the IEEE International Symposium on Industrial Electronics*, 2016, p.678-683 (modeling and DMPC of power networks)

[WBAL16a] B. Watkins, F. Berkel, S. Al-Areqi, S. Liu: Distributed Control of Constrained Linear Systems using a Resource Aware Communication Strategy for Networks with MAC. In: *Proceedings of the IEEE Multi-Conference on Systems and Control*, 2016, p.149-154 (event-triggered linear distributed control subject to constraints)

[WBAL16b] B. Watkins, F. Berkel, S. Al-Areqi, S. Liu: Event-based distributed control of dynamically coupled and constrained linear systems. In: *Proceedings of the 55th IEEE Conference on Decision and Control*, 2016, p. 6074-6079 (event-triggered linear distributed control subject to constraints)

[BWLG17] F. Berkel, B. Watkins, S. Liu, D. Görges: Output-Based Event-Triggered Model Predictive Control for Networked Control Systems. In: *Proceedings of the 20th IFAC World Congress*, 2017, p. 9281-9286 (centralized event-triggered output-based MPC; Chapter 4)

[BBL17] M. Bell, F. Berkel, S. Liu: Optimal Distributed Balancing Control for Three-Phase Four-Wire Low Voltage Grids. In: *Proceedings of the IEEE International Conference on Smart Grid Communications*, p.229-234, 2017 (DMPC of power networks)

[BL18c] F. Berkel, S. Liu: Non-Iterative Distributed Model Predictive Control with Event-Triggered Communication. In: *Proceedings of the American Control Conference*, 2018, p. 2344-2349 (DMPC with event-triggered communication; Chapter 5)

[BBBL18] F. Berkel, J. Bleich, M. Bell, S. Liu: A Distributed Voltage Controller for Medium Voltage Grids with Storage-containing Loads. In: *Proceedings of the 44th Annual Conference of the IEEE Industrial Electronics Society*, 2018 (DMPC of power networks with event-triggered communication; Chapter 7)

[BCBL18] F. Berkel, S. Caba, J. Bleich, S. Liu: A Modeling and Distributed MPC Approach for Water Distribution Networks. *Control Engineering Practice*, 81:199-206, 2018 (modeling and DMPC of water distribution networks; Chapter 7)

[BL18a] F. Berkel, S. Liu: An Event-triggered Cooperation Approach for Robust Distributed Model Predictive Control. *IFAC Journal of Systems and Control*, 6:16-24, 2018 (event-triggered DMPC; Chapter 6)

[BL18b] F. Berkel, S. Liu: An Event-triggered Output-based Model Predictive Control Strategy. *Accepted for publication in IEEE Transactions on Control of Network Systems*, 2018 (decentralized event-triggered output-based MPC; Chapter 4)

[BBL19] M. Bell, F. Berkel, S. Liu: Real-Time Distributed Control of Low Voltage Grids with Dynamic Optimal Power Dispatch of Renewable Energy Sources. *IEEE Transactions on Sustainable Energy*, 10(1):417-425, 2019. (modeling and DMPC of power networks; modeling in Chapter 7)

Assumption 2.2.1 implies that a linear control law $\boldsymbol{u}(\boldsymbol{x}) = \boldsymbol{K}\boldsymbol{x}$ with controller gain $\boldsymbol{K} \in \mathbb{R}^{m\times n}$ exists such that all eigenvalues of the matrix $\boldsymbol{A}+\boldsymbol{B}\boldsymbol{K}$ lie strictly inside the unit circle. Analogously, Assumption 2.2.2 implies that there exists an observer gain $\boldsymbol{L} \in \mathbb{R}^{n\times p}$ such that all eigenvalues of the matrix $\boldsymbol{A}-\boldsymbol{L}\boldsymbol{C}$ lie strictly inside the unit circle.

Linear Periodically Time-varying Systems

A linear time-varying discrete-time system is of the form

$$\mathscr{S} : \boldsymbol{x}_{t+1} = \boldsymbol{A}_t\boldsymbol{x}_t + \boldsymbol{B}_t\boldsymbol{u}_t + \boldsymbol{d}_t \tag{2.5}$$

$t \in \mathbb{N}$. The matrices $\boldsymbol{A}_t \in \mathbb{R}^{n\times n}$ and $\boldsymbol{B}_t \in \mathbb{R}^{n\times m}$ are assumed to be known at every time step. The vector $\boldsymbol{d}_t \in \mathbb{R}^n$ represents a known and bounded (possibly time-varying) exogenous input. The state vector $\boldsymbol{x}_t$ is measured and thus known.

Definition 2.2.3 (LPTV Systems). *The time-varying system* (2.5) *is called P-periodic if there exists a $P \in \mathbb{N}$ such that for all $t \in \mathbb{N}$, $\boldsymbol{A}_t = \boldsymbol{A}_{t+P}$, $\boldsymbol{B}_t = \boldsymbol{B}_{t+P}$ and $\boldsymbol{d}_t = \boldsymbol{d}_{t+P}$. The smallest P is called period of system* (2.5).

System (2.5) is considered as an LPTV system from now on.

The state and input of the system are subject to the (possibly time-varying) constraint

$$(\boldsymbol{x}_t, \boldsymbol{u}_t) \in \mathcal{Z}_t \tag{2.6}$$

for all $t \in \mathbb{N}$. The constraint sets $\mathcal{Z}_t \subseteq \mathbb{R}^{n+m}$ are compact and periodically time-varying with period P, i.e., $\mathcal{Z}_t = \mathcal{Z}_{t+P}\ \forall t \in \mathbb{N}$. Note that the constraints of the form (2.6) are also referred to as a mixed (state and input) constraints.

LPTV systems can be used for modeling of WDNs and power networks and are therefore relevant in Chapter 7. The system and input matrices $\boldsymbol{A}_t$ and $\boldsymbol{B}_t$ are usually time-invariant in these examples. However, an uncontrollable, time-varying but periodic demand which can be modeled as the periodic exogenous input $\boldsymbol{d}_t$ has to be satisfied.

2.2.2 Invariance and Lyapunov Stability

The definitions and theorems listed below are taken from [RM09, Appendix A&B] for LTI and [Gro15] as well as [LPM+14] for LPTV systems.

Definition 2.2.4 (Positive Invariant (PI) set). *A set $\mathcal{X} \subseteq \mathbb{R}^n$ is called a PI set for the autonomous closed-loop system $\boldsymbol{x}_{t+1} = \boldsymbol{A}\boldsymbol{x}_t + \boldsymbol{B}\boldsymbol{u}(\boldsymbol{x}_t)$, $t \in \mathbb{N}$ if $\boldsymbol{x}_t \in \mathcal{X}$ implies $\boldsymbol{x}_{t+1} \in \mathcal{X}$.*

Definition 2.2.5 (Robustly Positive Invariant (RPI) set). *A set $\mathcal{X} \subseteq \mathbb{R}^n$ is called an RPI set for the closed-loop system $\boldsymbol{x}_{t+1} = \boldsymbol{A}\boldsymbol{x}_t + \boldsymbol{B}\boldsymbol{u}(\boldsymbol{x}_t) + \boldsymbol{w}_t$ if $\boldsymbol{x}_t \in \mathcal{X}$ and $\boldsymbol{w}_t \in \mathcal{W}$ imply that $\boldsymbol{x}_{t+1} \in \mathcal{X}$.*

Definition 2.2.6 (Stable Set). *Suppose the sets $\mathcal{X} \subseteq \mathbb{R}^n$ and $\mathcal{A} \subseteq \mathbb{R}^n$ are PI for $\boldsymbol{x}_{t+1} = \boldsymbol{A}\boldsymbol{x}_t + \boldsymbol{B}\boldsymbol{u}_t(\boldsymbol{x}_t)$ and $\mathcal{A}$ is closed and lies in the interior of $\mathcal{X}$. The set $\mathcal{A}$ is said to be stable within $\mathcal{X}$ if for any $\epsilon \in \mathbb{R}_{>0}$ there exists a $\delta \in \mathbb{R}_{>0}$ such that $\boldsymbol{x}_0 \in \mathcal{X} \cap (\mathcal{A} \oplus \mathcal{B}_\delta(\boldsymbol{0}))$ implies that $\|\boldsymbol{x}_t\|_\mathcal{A} \leq \epsilon$ for all $t \in \mathbb{N}$.*

Definition 2.2.7 (Attractive Set). *Suppose the sets $\mathcal{X} \subseteq \mathbb{R}^n$ and $\mathcal{A} \subseteq \mathbb{R}^n$ are PI for $\boldsymbol{x}_{t+1} = \boldsymbol{A}\boldsymbol{x}_t + \boldsymbol{B}\boldsymbol{u}_t(\boldsymbol{x}_t)$ and $\mathcal{A}$ is closed and lies in the interior of $\mathcal{X}$. The set $\mathcal{A}$ is said to be attractive within $\mathcal{X}$ if $\lim\limits_{t\to\infty} \|\boldsymbol{x}_t\|_\mathcal{A} = 0$.*

Definition 2.2.8 (Globally Asymptotically Stable (GAS)). *Suppose the sets $\mathcal{X} \subseteq \mathbb{R}^n$ and $\mathcal{A} \subseteq \mathbb{R}^n$ are PI for $\boldsymbol{x}_{t+1} = \boldsymbol{A}\boldsymbol{x}_t + \boldsymbol{B}\boldsymbol{u}_t(\boldsymbol{x}_t)$ and $\mathcal{A}$ is closed and lies in the interior of $\mathcal{X}$. The set $\mathcal{A}$ is said to be GAS with Region Of Attraction (ROA) if it is stable and attractive.*

Definition 2.2.9 (Globally Exponentially Stable (GES)). *Suppose the sets $\mathcal{X} \subseteq \mathbb{R}^n$ and $\mathcal{A} \subseteq \mathbb{R}^n$ are PI for $\boldsymbol{x}_{t+1} = \boldsymbol{A}\boldsymbol{x}_t + \boldsymbol{B}\boldsymbol{u}_t(\boldsymbol{x}_t)$ and $\mathcal{A}$ is closed and lies in the interior of $\mathcal{X}$. The set $\mathcal{A}$ is said to be GES with ROA $\mathcal{X}$, if there exists a $c \in \mathbb{R}_{>0}$ and a $\gamma \in (0,1)$ such that $\|\boldsymbol{x}_t\|_\mathcal{A} \leq c\gamma^t \|\boldsymbol{x}_0\|_\mathcal{A}$ for all $\boldsymbol{x}_0 \in \mathcal{X}$, $t \in \mathbb{N}$.*

Definition 2.2.10 (Lyapunov function). *Suppose the sets $\mathcal{X} \subseteq \mathbb{R}^n$ and $\mathcal{A} \subseteq \mathbb{R}^n$ are PI for $\boldsymbol{x}_{t+1} = \boldsymbol{A}\boldsymbol{x}_t + \boldsymbol{B}\boldsymbol{u}_t(\boldsymbol{x}_t)$ and $\mathcal{A}$ is closed and lies in the interior of $\mathcal{X}$. A function $V : \mathbb{R}^n \to \mathbb{R}_{\geq 0}$ is said to be a Lyapunov function in $\mathcal{X}$ for the system $\boldsymbol{x}_{t+1} = \boldsymbol{A}\boldsymbol{x}_t + \boldsymbol{B}\boldsymbol{u}_t(\boldsymbol{x}_t)$ and the set $\mathcal{A}$ if there exist $\mathcal{K}_\infty$-class functions α_i with $i \in \mathbb{Z}_{1:2}$ and a $\mathcal{PD}$-class function α_3 such that for any $\boldsymbol{x}_t \in \mathcal{X}$*

$$\alpha_1(\|\boldsymbol{x}_t\|_\mathcal{A}) \leq V(\boldsymbol{x}_t) \leq \alpha_2(\|\boldsymbol{x}_t\|_\mathcal{A}) \tag{2.7}$$

$$V(\boldsymbol{A}\boldsymbol{x}_t + \boldsymbol{B}\boldsymbol{u}_t(\boldsymbol{x}_t)) - V(\boldsymbol{x}_t) \leq -\alpha_3(\|\boldsymbol{x}_t\|_\mathcal{A}). \tag{2.8}$$

Theorem 2.2.11 (Lyapunov function for GES). *Suppose the sets $\mathcal{X} \subseteq \mathbb{R}^n$ and $\mathcal{A} \subseteq \mathbb{R}^n$ are PI for $\boldsymbol{x}_{t+1} = \boldsymbol{A}\boldsymbol{x}_t + \boldsymbol{B}\boldsymbol{u}_t(\boldsymbol{x}_t)$ and $\mathcal{A}$ is closed and lies in the interior of $\mathcal{X}$. If there exists a Lyapunov function $V : \mathbb{R}^n \to \mathbb{R}_{\geq 0}$ with $\alpha_i(r) = a_i r^\sigma$ for $i \in \mathbb{Z}_{1:3}$ with $a_1, a_2, a_3, \sigma \in \mathbb{R}_{>0}$, then $\mathcal{A}$ is GES for $\boldsymbol{x}_{t+1} = \boldsymbol{A}\boldsymbol{x}_t + \boldsymbol{B}\boldsymbol{u}_t(\boldsymbol{x}_t)$ with ROA $\mathcal{X}$.*

The definitions and theorem above also hold for the special case of $\mathcal{A} = \{\boldsymbol{0}\}$, i.e., the case where the origin should be stabilized.

Theorem 2.2.12 (Stability of autonomous linear systems). *Consider the autonomous discrete-time LTI system $\boldsymbol{x}_{t+1} = \boldsymbol{A}\boldsymbol{x}_t$ with $t \in \mathbb{N}$ and $\boldsymbol{x}_t \in \mathbb{R}^n$. The origin is GAS if and only if the eigenvalues of the matrix $\boldsymbol{A}$ lie strictly inside the unit circle.*

The system matrix $\boldsymbol{A}$ of a system with GAS origin is also called a stable matrix.

Definition 2.2.13 (Periodically Positive Invariant (PPI) sets). *The sequence of sets $\mathcal{S}_r^{\mathrm{P}} \subseteq \mathbb{R}^n$ with $r \in \mathbb{Z}_{0:P-1}$ is called PPI for an LPTV system of the form $\boldsymbol{x}_{t+1} = \boldsymbol{A}_t\boldsymbol{x}_t + \boldsymbol{B}_t\boldsymbol{u}_t(\boldsymbol{x}_t) + \boldsymbol{d}_t$ if for each $t \in \mathbb{N}$ it holds that $\boldsymbol{x}_t \in \mathcal{S}_{[t]_P}^{\mathrm{P}}$ implies $\boldsymbol{x}_{t+1} \in \mathcal{S}_{[t+1]_P}^{\mathrm{P}}$.*

Definition 2.2.14 (Asymptotic stability). *Let the sequence of sets $\mathcal{S}_r^{\mathrm{P}} \subseteq \mathbb{R}^n$ with $r \in \mathbb{Z}_{0:P-1}$ be PPI for the LPTV system of the form $\boldsymbol{x}_{t+1} = \boldsymbol{A}_t\boldsymbol{x}_t + \boldsymbol{B}_t\boldsymbol{u}_t(\boldsymbol{x}_t) + \boldsymbol{d}_t$. Let the sequence $\boldsymbol{X}^{\mathrm{P}} = \left[\boldsymbol{x}_0^{\mathrm{P,T}}, \ldots, \boldsymbol{x}_{P-1}^{\mathrm{P,T}}\right]^{\mathrm{T}}$ be such that $\boldsymbol{x}_r^{\mathrm{P}} \in \mathcal{S}_r^{\mathrm{P}}$ for all $t \in \mathbb{Z}_{0:P-1}$. The sequence $\boldsymbol{X}^{\mathrm{P}}$ is said to be asymptotically stable if there exists a $\mathcal{KL}$-class function β such that $\left\|\boldsymbol{x}_t - \boldsymbol{x}_{[t]_P}^{\mathrm{P}}\right\| \leq \beta\left(\left\|\boldsymbol{x}_0 - \boldsymbol{x}_0^{\mathrm{P}}\right\|, t\right)$ for all $t \in \mathbb{N}$ with $\boldsymbol{x}_0 \in \mathcal{S}_0^{\mathrm{P}}$.*

2.3 Model Predictive Control

MPC is a powerful framework for the control of systems subject to constraints. The key idea is to solve at each time instant an optimal control problem over a finite horizon N using a model of the plant while minimizing some cost function. Constraints on the pant can be explicitly included into the optimization problem. As a result, a future input and state sequence is obtained. The first element of the input sequence is applied to the plant. The procedure is repeated at the next sampling instant. The principle is illustrated in Figure 2.1 for the case of nominal MPC, i.e., the case of no disturbance in the system. Due to the absence of disturbance, the predicted values $\boldsymbol{x}_{t+1|t}$ are identical to the states $\boldsymbol{x}_{t+1}$. For a more detailed introduction it is referred to the textbooks [Mac02, RM09, GP11, CB13].

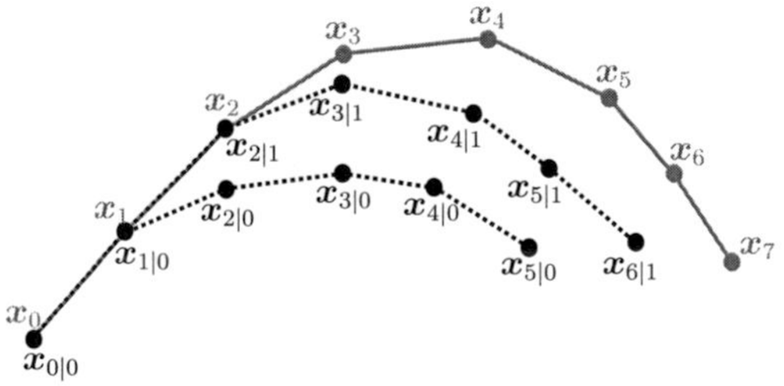

Figure 2.1: The figure illustrates the concept of nominal MPC for an example with prediction horizon $N = 5$. In dark gray the closed-loop trajectory $\boldsymbol{x}_t$ is shown. The open-loop predictions $\boldsymbol{x}_{t+r|t}$ for $r \in \mathbb{Z}_{0:N}$ at time $t = 0$ and $t = 1$ are depicted as black dotted lines.

The sequel of the section deals with robust MPC of LTI systems and economic MPC for LPTV systems.

The robust MPC schemes are relevant in this thesis for the event-triggered schemes in Chapter 4-6 where neglected or simplified modeling of the coupling between the subsystems and the event-triggered communication/cooperation style introduce uncertainties in the system. To prove recursive feasibility and convergence in those contexts, concepts from robust MPC are applied.

Economic MPC concepts for LPTV systems are relevant for the methods presented in Chapter 7. Here infrastructure systems are considered where an economically motivated cost function which is in general not positive-definite to some reference is minimized. Concepts known from economic MPC can be used to give guarantees on the resulting behavior of the closed-loop system.

2.3.1 Robust Model Predictive Control

When the system under control is subject to uncertainty, the predicted state $\boldsymbol{x}_{t+1|t}$ and the plant's state $\boldsymbol{x}_{t+1}$ usually differ and feasibility as well as stability of the nominal MPC can get lost. The uncertainty can stem from different reasons, such as external disturbances, a mismatch between model and actual plant or measurement noise. In robust MPC the main idea is to change the problem formulation of nominal MPC to recover the desired closed-loop properties. A variety of possible robust approaches has been proposed in the literature. In [May14] the approaches are categorized into three groups, which are shortly reviewed in the sequel.

The first option is to exploit the inherent robustness property of the nominal MPC controller. For some special cases, it can be shown that a system with disturbance that is controlled by a model predictive controller designed for the nominal system is input-to-state stable. A detailed treatment of the topic can be found in [RM09, chapter 3.2].

In the second group, the disturbance is taken into account in the online control problem. The state and control constraints are required to be satisfied for all possible disturbance sequences. In [LAC02] a method for ensuring recursive feasibility, using an RPI terminal set and a terminal penalty, is presented. The nominal cost function is minimized. In [KM04, LASC06, LMHA08] the problem of min-max MPC is addressed, i.e., a min-max cost function is considered. These papers employ optimization over control policies rather than control sequences. This methodology is theoretically preferable but the decision variable is infinite dimensional. Hence, its computational complexity inhibits its practical use.

The third group consists of the tube-based MPC methods. The key idea of them is to ensure that the closed-loop trajectories lie in a tube around the open-loop predictions. The most popular tube-based methods are probably the ones proposed in [CRZ01, MSR05].

In this thesis the method of [CRZ01] is employed due to its convenient inclusion into the event-triggering framework. The method is shortly reviewed in the sequel of this subsection. The presentation partly follows [CRZ01], [RM09] and [BHA17].

The approach in [CRZ01] considers state-feedback LTI systems of the form (2.1), i.e.,

$$\boldsymbol{x}_{t+1} = \boldsymbol{A}\boldsymbol{x}_t + \boldsymbol{B}\boldsymbol{u}_t + \boldsymbol{w}_t \tag{2.9}$$

for $t \in \mathbb{N}$ where $\boldsymbol{w}_t \in \mathcal{W}$ and $\mathcal{W}$ is a compact set. The control parametrization $\boldsymbol{u}_t = \boldsymbol{v}_t + \boldsymbol{K}\boldsymbol{x}_t$ is used where $\boldsymbol{v}_t \in \mathbb{R}^m$ and the controller gain $\boldsymbol{K} \in \mathbb{R}^{m \times n}$ is chosen such that the following main assumption holds.

Assumption 2.3.1. *The matrix* $\boldsymbol{F} = \boldsymbol{A} + \boldsymbol{B}\boldsymbol{K}$ *is stable.*

Considering the nominal system of (2.1) and the controller parametrization one obtains

$$\bar{\boldsymbol{x}}_{t+1} = \boldsymbol{F}\bar{\boldsymbol{x}}_t + \boldsymbol{B}\boldsymbol{v}_t \quad t \in \mathbb{N}. \tag{2.10}$$

$\boldsymbol{v}$ can be thought of as the input of system (2.10). The difference between the actual state $\boldsymbol{x}_t$ and the nominal state $\bar{\boldsymbol{x}}_t$ satisfies the difference equation

$$\boldsymbol{x}_{t+1} - \bar{\boldsymbol{x}}_{t+1} = \boldsymbol{F}\left(\boldsymbol{x}_t - \bar{\boldsymbol{x}}_t\right) + \boldsymbol{w}_t \quad t \in \mathbb{N}. \tag{2.11}$$

Thus, if $\boldsymbol{x}_0 = \bar{\boldsymbol{x}}_0$, then

$$\boldsymbol{x}_t \in \{\bar{\boldsymbol{x}}_t\} \oplus \mathcal{F}^t \tag{2.12}$$

$$\text{where } \mathcal{F}^t = \bigoplus_{r=0}^{t-1} \boldsymbol{F}^r \mathcal{W} \tag{2.13}$$

$\forall t \in \mathbb{N}$. Moreover, $\boldsymbol{u}_t \in \{\bar{\boldsymbol{u}}_t\} \oplus \boldsymbol{K}\mathcal{F}^t$ where $\bar{\boldsymbol{u}}_t = \boldsymbol{v}_t + \boldsymbol{K}\bar{\boldsymbol{x}}_t$. Loosely speaking, this means that the state $\boldsymbol{x}_t$ and input $\boldsymbol{u}_t$ are in the proximity of the references $\bar{\boldsymbol{x}}_t$ and $\bar{\boldsymbol{u}}_t$. If the constraints of the nominal problem are tightened properly, the constraints on the states and inputs can be satisfied.

The online optimal control problem is stated as

$$\begin{aligned}
\underset{\boldsymbol{V}_t}{\text{minimize}} \quad & J_N(\boldsymbol{V}_t) = \sum_{r=t}^{t+N-1} l(\boldsymbol{v}_{r|t}) && \text{(2.14a)}\\
\text{subject to} \quad & \bar{\boldsymbol{x}}_{t|t} = \boldsymbol{x}_t, && \text{(2.14b)}\\
& \bar{\boldsymbol{x}}_{t+N|t} \in \mathcal{X}^{\mathrm{f}}, && \text{(2.14c)}\\
& \text{and } \forall r \in \mathbb{Z}_{0:N-1}: && \text{(2.14d)}\\
& \bar{\boldsymbol{x}}_{t+r+1|t} = \boldsymbol{F}\bar{\boldsymbol{x}}_{t+r|t} + \boldsymbol{B}\boldsymbol{v}_{t+r|t}, && \text{(2.14e)}\\
& \bar{\boldsymbol{x}}_{t+r|t} \in \bar{\mathcal{X}}^r, && \text{(2.14f)}\\
& \boldsymbol{v}_{t+r|t} + \boldsymbol{K}\bar{\boldsymbol{x}}_{t+r|t} \in \bar{\mathcal{U}}^r. && \text{(2.14g)}
\end{aligned}$$

The index $t+r|t$ denotes a predicted value for time $t+r$ at time t with $r \in \mathbb{Z}_{0:N}$. $N \in \mathbb{N}_{\geq 1}$ is the prediction horizon. The cost function J_N is the sum of the stage cost $l : \mathbb{R}^m \to \mathbb{R}_{\geq 0}$ which is assumed to be a $\mathcal{PD}$-class function. The sequence $\boldsymbol{V}_t = \left[\boldsymbol{v}_{t|t}^{\mathrm{T}}, \ldots, \boldsymbol{v}_{t+N-1|t}^{\mathrm{T}}\right]^{\mathrm{T}}$ consists of the predicted value $\boldsymbol{v}_{l|t}$ with $l \in \mathbb{Z}_{t:t+N-1}$.

The tightened constraint sets $\bar{\mathcal{X}}^r \subseteq \mathbb{R}^n$ and $\bar{\mathcal{U}}^r \subseteq \mathbb{R}^m$ are given by

$$\begin{aligned}
\bar{\mathcal{X}}^r &= \mathcal{X} \ominus \mathcal{F}^r && \forall r \in \mathbb{Z}_{0:N}, && \text{(2.15a)}\\
\bar{\mathcal{U}}^r &= \mathcal{U} \ominus \boldsymbol{K}\mathcal{F}^r && \forall r \in \mathbb{Z}_{0:N}. && \text{(2.15b)}
\end{aligned}$$

In Figure 2.2 the principle of the constraint tightening approach is illustrated. The terminal set $\mathcal{X}^{\mathrm{f}} \subseteq \mathbb{R}^n$ has to satisfy the following assumption.

Assumption 2.3.2. *There exists a non-empty set $\mathcal{X}^{\mathrm{f}}$ such that*

$$\boldsymbol{K}\mathcal{X}^{\mathrm{f}} \subseteq \bar{\mathcal{U}}^N, \quad \mathcal{X}^{\mathrm{f}} \subseteq \bar{\mathcal{X}}^N, \quad \boldsymbol{F}\mathcal{X}^{\mathrm{f}} \oplus \boldsymbol{F}^N\mathcal{W} \subseteq \mathcal{X}^{\mathrm{f}}.$$

The assumption requires that $\mathcal{X}^{\mathrm{f}}$ is an RPI set which is a standard assumption in many robust MPC approaches (see e.g. [CRZ01, LAC02]). It is important to note that for large sets $\mathcal{W}$ the sets $\bar{\mathcal{X}}^N$ and $\bar{\mathcal{U}}^N$ can become empty and thus the design can fail.

Before the main properties can be stated, the following definitions are required. The set of admissible sequences $\boldsymbol{V}$ is defined as

$$\mathcal{D}_N(\boldsymbol{x}) = \left\{\boldsymbol{V} \in \mathbb{R}^{m \cdot N} | (2.14) \text{ is feasible}\right\}. \tag{2.16}$$

Furthermore, the set of states for which a sequence $\boldsymbol{V}$ can be found such that all constraints are feasible is defined as

$$\mathcal{X}_N = \{\boldsymbol{x} \in \mathbb{R}^n | \mathcal{D}_N(\boldsymbol{x}) \neq \emptyset\}. \tag{2.17}$$

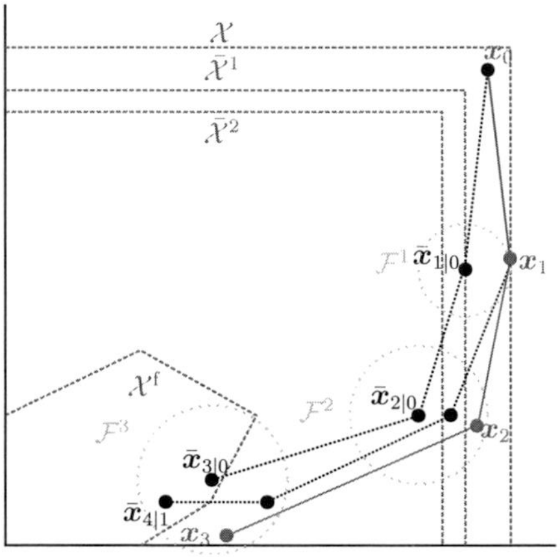

Figure 2.2: The figure illustrates the principle of tubes and the constraint tightening approach for an example with prediction horizon $N = 3$. The closed-loop state trajectory $\boldsymbol{x}_t$ is shown in dark gray. The open-loop state predictions $\bar{\boldsymbol{x}}_{t+r|t}$ for $r \in \mathbb{Z}_{0:N}$ are shown as dotted black lines for $t \in \{0, 1\}$. The sets $\mathcal{F}^r$ are depicted as dotted light gray circles. One can see that the state trajectory lies in the proximity of the open-loop prediction for the case $t = 0$ (cf. equation (2.12)). In dark gray dashed lines the tightened state constraints $\bar{\mathcal{X}}^1$ and $\bar{\mathcal{X}}^2$ are drawn.

Another important set for analysis purposes is

$$\mathcal{X}_K = \left\{\boldsymbol{x} \in \mathbb{R}^n | \boldsymbol{F}^N \boldsymbol{x} \in \mathcal{X}^{\mathrm{f}}, \boldsymbol{F}^l \boldsymbol{x} \in \bar{\mathcal{X}}^l, \boldsymbol{K}\boldsymbol{F}^l \boldsymbol{x} \in \bar{\mathcal{U}}^l \ \forall l \in \mathbb{Z}_{0:N-1}\right\}. \tag{2.18}$$

The input applied to the system is

$$\boldsymbol{u}_t\left(\boldsymbol{x}_t\right) = \boldsymbol{v}_t + \boldsymbol{K}\boldsymbol{x}_t \tag{2.19}$$

where $\boldsymbol{v}_t = \boldsymbol{v}^*_{t|t}\left(\boldsymbol{x}_t\right)$ is the first element of the optimal sequence

$$\boldsymbol{V}^*_t\left(\boldsymbol{x}_t\right) = \arg\min_{\boldsymbol{V}_t \in \mathcal{D}_N(\boldsymbol{x}_t)} J_N\left(\boldsymbol{V}_t\right). \tag{2.20}$$

The optimal cost function is $J^*_N\left(\boldsymbol{x}_t\right) = \min_{\boldsymbol{V}_t \in \mathcal{D}_N(\boldsymbol{x}_t)} J_N\left(\boldsymbol{V}_t\right)$. For the closed-loop the following important properties hold.

Lemma 2.3.3. *Consider system* (2.9) *under control law* (2.19) *and assume* $\boldsymbol{x}_0 \in \mathcal{X}_N$*, then* $\boldsymbol{x}_t \in \mathcal{X}_N$ *for all* $t \in \mathbb{N}$ *and for the optimal cost function*

$$J_N^*(\boldsymbol{x}_{t+1}) - J_N^*(\boldsymbol{x}_t) \leq -l(\boldsymbol{v}_t) \tag{2.21}$$

$\forall t \in \mathbb{N}$ *holds.*

Proof. (Sketch) The proofs for both statements are given in [CRZ01, Lemma 7 and Theorem 8]. However, the proof of (2.21) is sketched here as the idea is repeatedly applied in the thesis.

Assuming an optimal sequence $\boldsymbol{V}_t^* = \left[\boldsymbol{v}_{t|t}^{*,\mathrm{T}}, \ldots, \boldsymbol{v}_{t+N-1|t}^{*,\mathrm{T}}\right]^{\mathrm{T}}$ is given at time t, a feasible sequence $\boldsymbol{V}_{t+1}^{\text{shift}}$ for time $t+1$ can be generated by shifting the sequence $\boldsymbol{V}_t^*$ and extending it with zero, i.e.,

$$\boldsymbol{V}_{t+1}^{\text{shift}} = \left[\boldsymbol{v}_{t+1|t}^{*,\mathrm{T}}, \ldots, \boldsymbol{v}_{t+N-1|t}^{*,\mathrm{T}}, \boldsymbol{0}^{\mathrm{T}}\right]^{\mathrm{T}}. \tag{2.22}$$

Evaluating the cost for the sequence $\boldsymbol{V}_{t+1}^{\text{shift}}$ leads to

$$J_N\left(\boldsymbol{V}_{t+1}^{\text{shift}}\right) = J_N^*(\boldsymbol{x}_t) - l(\boldsymbol{v}_t). \tag{2.23}$$

By optimality, $J_N^*(\boldsymbol{x}_{t+1}) \leq J_N\left(\boldsymbol{V}_{t+1}^{\text{shift}}\right)$ and, thus, it follows that (2.21) holds. □

The first statement, $\boldsymbol{x}_0 \in \mathcal{X}_N$ implies that $\boldsymbol{x}_t \in \mathcal{X}_N$ for all $t \in \mathbb{N}$, says that the $\mathcal{X}_N$ is an RPI set. This property is also known as recursive feasibility in the literature.

The next theorem summarizes the stability properties of the robust MPC approach.

Theorem 2.3.4. *Consider system* (2.9) *with control law* (2.19) *and assume* $\boldsymbol{x}_0 \in \mathcal{X}_N$*, then the state* $\boldsymbol{x}_t$ *converges to the set* $\mathcal{F}^\infty$*, i.e.,* $\lim_{t\to\infty} \|\boldsymbol{x}_t\|_{\mathcal{F}^\infty} = 0$*. If there exists an* $\epsilon \in \mathbb{R}_{>0}$ *such that* $\mathcal{F}^\infty \oplus \mathcal{B}_\epsilon(\boldsymbol{0}) \subseteq \mathcal{X}_K$*, then the set* $\mathcal{F}^\infty$ *is GAS with ROA* $\mathcal{X}_N$*.*

The first part of the proof can be found in [CRZ01, Theorem 8], the second part follows the lines of [BHA17, Theorem 2]. It is shown in [KG98] that the set $\mathcal{F}^\infty$ is an RPI set for the dymanics (2.11). In fact it is the set for which every other RPI set $\mathcal{S}$ satisfies $\mathcal{S} \supseteq \mathcal{F}^\infty$. Hence, $\mathcal{F}^\infty$ is also called the minimal RPI set of (2.11).

Remark 2.3.5. *In [CRZ01] the quadratic stage cost* $l(\boldsymbol{v}) = \boldsymbol{v}^{\mathrm{T}}\boldsymbol{M}\boldsymbol{v}$ *with* $\boldsymbol{M} \in \mathbb{R}^{m\times m}$*,* $\boldsymbol{M} \succ \boldsymbol{0}$ *is used. Moreover, it is suggested to select* $\boldsymbol{M} = \boldsymbol{R} + \boldsymbol{B}^{\mathrm{T}}\boldsymbol{P}\boldsymbol{B}$ *where* $\boldsymbol{P}$ *is the solution of the discrete-time algebraic Riccati equation* $\boldsymbol{P} = \boldsymbol{Q} + \boldsymbol{A}^{\mathrm{T}}\boldsymbol{P}\boldsymbol{A} - \boldsymbol{A}^{\mathrm{T}}\boldsymbol{P}\boldsymbol{B}\left(\boldsymbol{R} + \boldsymbol{B}^{\mathrm{T}}\boldsymbol{P}\boldsymbol{B}\right)^{-1}\boldsymbol{B}^{\mathrm{T}}\boldsymbol{P}\boldsymbol{A}$ *where* $\boldsymbol{Q} \in \mathbb{R}^{n\times n}$ *with* $\boldsymbol{Q} \succeq \boldsymbol{0}$ *and* $\boldsymbol{R} \in \mathbb{R}^{m\times m}$ *with* $\boldsymbol{R} \succ \boldsymbol{0}$ *are weighting matrices.*

Remark 2.3.6. *In the case that $\mathcal{X}$, $\mathcal{U}$ and $\mathcal{W}$ are polytopes, the sets $\bar{\mathcal{X}}^l$ and $\bar{\mathcal{U}}^l$ can be computed by methods presented in [BBM17, Chapter 4]. A polytopic terminal set $\mathcal{X}^\mathrm{f}$ satisfying Assumption 2.3.2 can be computed as shown in [BBM17, Chapter 10]. The number of halfspaces to describe the polytopes $\bar{\mathcal{X}}^l$ and $\bar{\mathcal{U}}^l$ is identical to the number of halfspaces to describe the original constraints $\mathcal{X}$, $\mathcal{U}$. This means that the computational complexity of the robust MPC problem is almost the same as of the nominal one. This implies that the computational effort of the approaches in this thesis which are relying on robust MPC methods have a similar computational complexity as their time-triggered equivalents.*

However, it should be noted that the above-mentioned methods for computing the polytopic sets can suffer from a high calculation effort even for small-scale problems as well as numerical problems.

Remark 2.3.7. *In the case that the sets $\bar{\mathcal{X}}^l$, $\bar{\mathcal{U}}^l$ and $\mathcal{X}^\mathrm{f}$ are polytopes, optimization problem* (2.14) *can be recast as a Quadratic Program (QP). If $\bar{\mathcal{X}}^l$ as well as $\bar{\mathcal{U}}^l$ are polytopes and $\mathcal{X}^\mathrm{f}$ is an ellipsoid,* (2.14) *can be formulated as a Quadratically Constrained Quadratic Program (QPQC). Efficient solvers exist for both cases.*

2.3.2 Economic Model Predictive Control for LPTV Systems

In this section the control of the LPTV system (2.5) subject to the constraint (2.6) is considered. The performance of the system is measured by the time-varying stage cost $l_t : \mathbb{R}^n \times \mathbb{R}^m \to \mathbb{R}$ which has to satisfy the following assumption.

Assumption 2.3.8. *The stage cost l_t is continuous and periodically time-varying, i.e., $l_t(\cdot,\cdot) = l_{t+P}(\cdot,\cdot)$.*

In general, the stage cost l_t does need not satisfy any strict convexity or positive definiteness assumptions with respect to a set-point. It ideally reflects the economic costs of the system under control.

To determine the optimal behavior of the system, the following optimization problem is solved offline:

$$\underset{\boldsymbol{x}_0,\boldsymbol{U}}{\text{minimize}} \quad \sum_{r=0}^{P-1} l_r(\boldsymbol{x}_r, \boldsymbol{u}_r) \tag{2.24a}$$

$$\text{subject to} \quad \boldsymbol{x}_P = \boldsymbol{x}_0 \tag{2.24b}$$

$$\text{for all } r \in \mathbb{Z}_{0:P-1}:$$

$$\boldsymbol{x}_{r+1} = \boldsymbol{A}_r \boldsymbol{x}_r + \boldsymbol{B}_r \boldsymbol{u}_r + \boldsymbol{d}_r \tag{2.24c}$$

$$(\boldsymbol{x}_r, \boldsymbol{u}_r) \in \mathcal{Z}_r. \tag{2.24d}$$

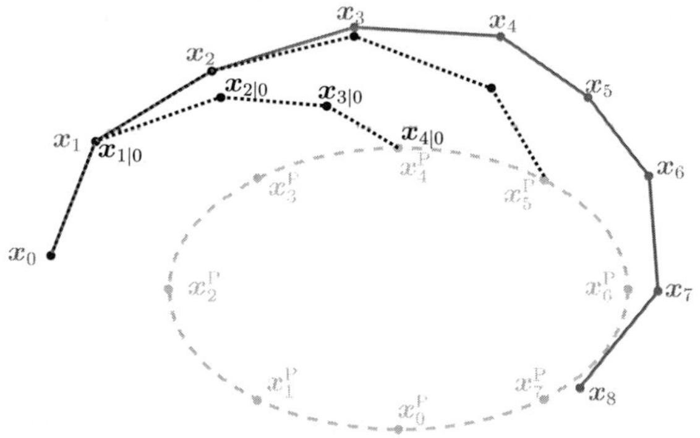

Figure 2.3: The figure illustrates the principle of the control scheme for an example with $P = 8$ and $N = 4$. In light gray the optimal sequence $\boldsymbol{X}^{\mathrm{P}}$ is shown. Furthermore, the closed-loop trajectory $\boldsymbol{x}_t$ is depicted in dark gray. The black dotted open-loop predictions $\boldsymbol{x}_{r|t}$ for $r \in \mathbb{Z}_{0:N}$ at time $t = 0$ and $t = 1$ are shown.

From the solution of the problem the sequences $\boldsymbol{X}^{\mathrm{P}}$ and $\boldsymbol{U}^{\mathrm{P}}$ which are called the optimal periodic state and input sequence, respectively, can be constructed. Constraint (2.24b) ensures periodicity of the solution.

The online optimal control problem is

$$\begin{aligned}
&\underset{\boldsymbol{U}_t}{\text{minimize}} && V_{N,t}(\boldsymbol{x}_t, \boldsymbol{U}_t) = \sum_{r=t}^{t+N-1} l_r\left(\boldsymbol{x}_{r|t}, \boldsymbol{u}_{r|t}\right) && (2.25\text{a})\\
&\text{subject to} && \boldsymbol{x}_{t|t} = \boldsymbol{x}_t && (2.25\text{b})\\
& && \boldsymbol{x}_{t+N|t} = \boldsymbol{x}^{\mathrm{P}}_{[t+N]_P} && (2.25\text{c})\\
& && \text{for all } r \in \mathbb{Z}_{t:t+N-1}: && \\
& && \boldsymbol{x}_{r+1|t} = \boldsymbol{A}_r \boldsymbol{x}_{r|t} + \boldsymbol{B}_r \boldsymbol{u}_{r|t} + \boldsymbol{d}_{r|t} && (2.25\text{d})\\
& && \left(\boldsymbol{x}_{r|t}, \boldsymbol{u}_{r|t}\right) \in \mathcal{Z}_r. && (2.25\text{e})
\end{aligned}$$

In constraint (2.25c) the optimal periodic state sequence is exploited as a terminal constraint. Figure 2.3 illustrates the concept of the periodic MPC scheme.

The input admissible set $\mathcal{D}_{N,t}$ is defined as

$$\mathcal{D}_{N,t}(\boldsymbol{x}) = \left\{\boldsymbol{U} \in \mathbb{R}^{m \cdot N} \,|\, (2.25) \text{ is feasible}\right\}. \tag{2.26}$$

The set of states for which an admissible input can be found is defined as

$$\mathcal{X}_{N,t} = \{\boldsymbol{x} \in \mathbb{R}^n | \mathcal{D}_{N,t}(\boldsymbol{x}) \neq \emptyset\}. \tag{2.27}$$

The input applied to the system is

$$\boldsymbol{u}_t(\boldsymbol{x}_t) = \boldsymbol{u}^*_{t|t}(\boldsymbol{x}_t) \tag{2.28}$$

where $\boldsymbol{u}^*_{t|t}$ is the first element of the optimal input sequence

$$\boldsymbol{U}^*_t(\boldsymbol{x}_t) = \arg\min_{\boldsymbol{U}_t \in \mathcal{D}_{N,t}(\boldsymbol{x}_t)} V_{N,t}(\boldsymbol{x}_t, \boldsymbol{U}_t). \tag{2.29}$$

Next, stability of the closed-loop is discussed. For this purpose the following assumptions are required.

Assumption 2.3.9 (Strict convexity). *The stage cost l_t is a strictly convex function. The sets $\mathcal{Z}_t$ are convex.*

Assumption 2.3.10 (Weak controllability). *There exists a $\mathcal{K}_\infty$-class function $\alpha(\cdot)$ such that for every $\boldsymbol{x}_t \in \mathcal{X}_{N,t}$ there exists a sequence $\boldsymbol{U}^{\mathrm{T}} = \left[\boldsymbol{u}_t^{\mathrm{T}}, \ldots, \boldsymbol{u}_{t+N-1}^{\mathrm{T}}\right] \in \mathcal{D}_{N,t}(\boldsymbol{x}_t)$ and*

$$\sum_{r=t}^{t+N-1} \left\|\boldsymbol{u}_r - \boldsymbol{u}^{\mathrm{P}}_{[r]_P}\right\| \leq \alpha\left(\left\|\boldsymbol{x}_t - \boldsymbol{x}^{\mathrm{P}}_{[t]_P}\right\|\right). \tag{2.30}$$

Assumption 2.3.9 ensures that (2.24) and (2.25) are strictly convex optimization problems for which strong duality holds. The assumption could be replaced by a less restrictive dissipativity assumption as in [GOMP17]. For the sake of simplicity this is not done in this thesis. Assumption 2.3.10 implies that the cost for steering the system from $\boldsymbol{x}_t$ to $\boldsymbol{x}^{\mathrm{P}}_{[t+N]_P}$ is bounded. Now the results on feasibility and stability can be established.

Theorem 2.3.11. *Let $\boldsymbol{X}^{\mathrm{P}}$ be given as the solution of* (2.24). *If $\boldsymbol{x}_0 \in \mathcal{X}_{N,0}$, then $\boldsymbol{x}_t \in \mathcal{X}_{N,t}$ for all $t \in \mathbb{N}$ and $\boldsymbol{X}^{\mathrm{P}}$ is an asymptotically stable sequence of the system* (2.5) *under the feedback* (2.28).

The proof can be found in [ZGD13].

Remark 2.3.12. *In optimization problem* (2.25) *the terminal equality constraint MPC formulation is used. It is well known that this formulation can be restrictive. In [GOMP17] the terminal equality constraint is replaced with a terminal penalty and a terminal region constraint. The formulation is, however, not required in this thesis.*

3 Modeling and Control of Coupled Linear Subsystems

In this section the problem set-ups considered in this thesis, decentralized control of NCSs and distributed control of coupled subsystems, are presented. The considered types of plant models and the model of the communication network are discussed. Furthermore, a suboptimal robust MPC approach which is applied in this thesis is introduced.

3.1 Problem Set-up

The general set-up considered throughout the thesis is that the overall plant $\mathscr{S}$ can be decomposed into M subsystems $\mathscr{S}_i$ which are coupled in cost, dynamics or constraints. Each subsystem is controlled by a local model predictive controller $\mathscr{C}_i$.

3.1.1 Decentralized Control of Networked Control Systems

The considered set-up of decentralized control of NCSs is depicted in Figure 3.1.

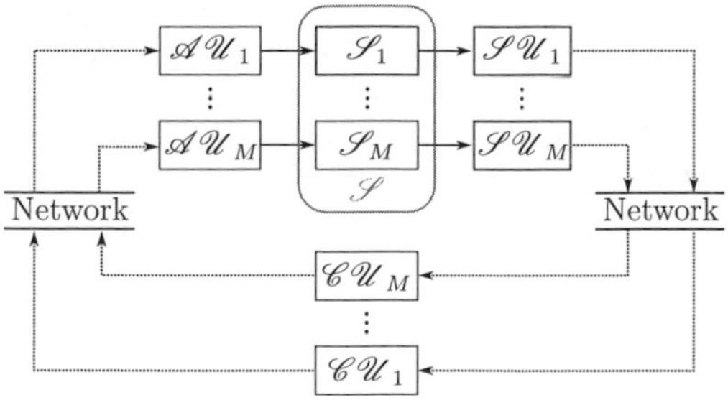

Figure 3.1: The figure shows the networked control configuration for coupled subsystems $\mathscr{S}_i$. The dashed lines are the communication paths over the network from sensor to controller and controller to actuator.

Each subsystem has its own control loop with a sensor unit $\mathscr{SU}_i$, a controller unit $\mathscr{CU}_i$ and an actuator unit $\mathscr{AU}_i$. The sensor measures the local state or output and transmits the measurement to the controller unit over the S2C channel of a communication network. The controller unit calculates a new input (sequence) and transmits it to the actuator unit over the C2A channel of the network where the input is then applied to the subsystem. There is no communication between the controllers. Thus, this is a decentralized scheme. As the network is shared, possibly wireless, and communication as well as energy resources are limited, it is desired to reduce the transmissions. The aim is to design sensor, controller and actuator units such that the transmissions over the network are kept low while constraints are satisfied and a desired closed-loop performance is achieved.

3.1.2 Distributed Control of Coupled Subsystems

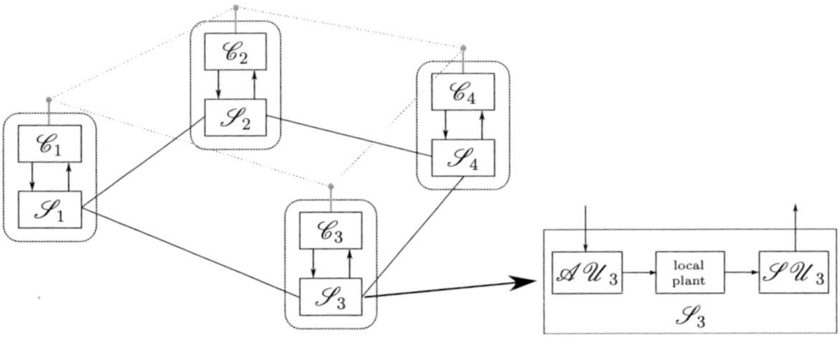

Figure 3.2: The figure shows an example set-up for distributed control of $M = 4$ coupled subsystems. The black solid lines indicate the coupling between the subsystems $\mathscr{S}_i$ which are controlled by a local controller $\mathscr{C}_i$. In dashed gray the communication network is depicted. On the right, it is zoomed into subsystem $\mathscr{S}_3$ and its components are shown.

The set-up for distributed control of coupled subsystems is shown in Figure 3.2. In each subsystem the local state is measured. It is assumed that the actuator and sensor units are integrated in the subsystem and can communicate directly with the controller. Therefore, the communication effort between them is not of interest and both units are no longer explicitly mentioned in this set-up, as it is the case in the majority of DMPC publications. Moreover, the controller units are simply called controller in this set-up. They can communicate with each other over a communication network in order to improve the overall performance of the system. The aim of this thesis is to develop controllers such that the communication between the controllers of the subsystems is kept low by communicating only when it is required to achieve the desired closed-loop performance.

3.2 Modeling of Coupled Linear Subsystems

The considered types of systems consist of coupled subsystems with linear time-invariant or periodically time-varying models. These are presented in the following section.

3.2.1 Linear Time-invariant Systems

The local dynamics of each subsystem $\mathscr{S}_i$, $i \in \mathbb{Z}_{1:M}$, are given by the LTI discrete-time model

$$\mathscr{S}_i : \begin{cases} \boldsymbol{x}_{i,t+1} & = \boldsymbol{A}_{ii}\boldsymbol{x}_{i,t} + \boldsymbol{B}_{ii}\boldsymbol{u}_{i,t} + \sum_{j\in\mathcal{N}_i} \boldsymbol{A}_{ij}\boldsymbol{x}_{j,t} + \boldsymbol{B}_{ij}\boldsymbol{u}_{j,t} + \boldsymbol{w}_{i,t} \\ \boldsymbol{y}_{i,t} & = \boldsymbol{C}_i\boldsymbol{x}_{i,t} \end{cases} \tag{3.1}$$

$t \in \mathbb{N}$. $\boldsymbol{x}_i \in \mathbb{R}^{n_i}$, $\boldsymbol{u}_i \in \mathbb{R}^{m_i}$ and $\boldsymbol{y}_i \in \mathbb{R}^{p_i}$ are the local state, input and output, respectively. $\boldsymbol{w}_i \in \mathbb{R}^{n_i}$ are unknown disturbances. They satisfy the constraints $\boldsymbol{w}_{i,t} \in \mathcal{W}_i$ for all $t \in \mathbb{N}$ where $\mathcal{W}_i$ are compact sets. $\boldsymbol{A}_{ii} \in \mathbb{R}^{n_i \times n_i}$, $\boldsymbol{B}_{ii} \in \mathbb{R}^{n_i \times m_i}$, $\boldsymbol{C}_i \in \mathbb{R}^{p_i \times n_i}$ are local system, input and output matrix, respectively. $\boldsymbol{A}_{ij} \in \mathbb{R}^{n_i \times n_j}$, $\boldsymbol{B}_{ij} \in \mathbb{R}^{n_i \times m_j}$ with $j \neq i$ are the coupling system and input matrices. $\mathcal{N}_i$ is an ordered set defined as $\mathcal{N}_i = \{j \in \mathbb{Z}_{1:M} \setminus \{i\} | \boldsymbol{A}_{ij} \neq \boldsymbol{0} \vee \boldsymbol{B}_{ij} \neq \boldsymbol{0}\}$. It is called the set of neighboring subsystems of $\mathscr{S}_i$.

The states and inputs are subject to local constraints

$$\boldsymbol{x}_{i,t} \in \mathcal{X}_i \subseteq \mathbb{R}^{n_i}, \quad \boldsymbol{u}_{i,t} \in \mathcal{U}_i \subseteq \mathbb{R}^{m_i}, \tag{3.2}$$

$t \in \mathbb{N}$ where $\mathcal{X}_i$ and $\mathcal{U}_i$ are compact convex sets containing the origin in their interior.

System (3.1) and the constraints (3.2) can be recast in the form (2.1) and (2.3). The global state, input and disturbance are $\boldsymbol{x} = \mathrm{col}_{i\in\mathbb{Z}_{1:M}}(\boldsymbol{x}_i)$, $\boldsymbol{u} = \mathrm{col}_{i\in\mathbb{Z}_{1:M}}(\boldsymbol{u}_i)$ and $\boldsymbol{w} = \mathrm{col}_{i\in\mathbb{Z}_{1:M}}(\boldsymbol{w}_i)$, respectively. The global system matrix $\boldsymbol{A}$ and the matrices $\boldsymbol{A}_{ij}$ are related by $\boldsymbol{A} = (\boldsymbol{A}_{ij})$. The input matrix $\boldsymbol{B}$ and the matrices $\boldsymbol{B}_{ij}$ are related by $\boldsymbol{B} = (\boldsymbol{B}_{ij})$. The global output matrix $\boldsymbol{C} = \mathrm{diag}_{i\in\mathbb{Z}_{1:M}}(\boldsymbol{C}_i)$. The global constraint sets in (2.3) are given by $\mathcal{U} = \prod_{i=1}^{M} \mathcal{U}_i$, $\mathcal{X} = \prod_{i=1}^{M} \mathcal{X}_i$.

3.2.2 Linear Periodically Time-varying Systems

The local dynamics of each subsystem $\mathscr{S}_i$, $i \in \mathbb{Z}_{1:M}$, are given by the LPTV discrete-time model

$$\mathscr{S}_i : \boldsymbol{x}_{i,t+1} = \boldsymbol{A}_{ii,t}\boldsymbol{x}_{i,t} + \boldsymbol{B}_{ii,t}\boldsymbol{u}_{i,t} + \sum_{j\in\mathcal{N}_i} \boldsymbol{B}_{ij,t}\boldsymbol{u}_{j,t} + \boldsymbol{d}_{i,t} \tag{3.3}$$

$t \in \mathbb{N}$. The local system matrix $\boldsymbol{A}_{ii,t} \in \mathbb{R}^{n_i\times n_i}$, local input matrix $\boldsymbol{B}_{ii,t} \in \mathbb{R}^{n_i\times m_i}$ and coupling input matrix $\boldsymbol{B}_{ij,t} \in \mathbb{R}^{n_i\times m_j}$ are periodically time-varying with period P, i.e., $\boldsymbol{A}_{ii,t} = \boldsymbol{A}_{ii,t+P}$, $\boldsymbol{B}_{ii,t} = \boldsymbol{B}_{ii,t+P}$ and $\boldsymbol{B}_{ij,t} = \boldsymbol{B}_{ij,t+P}$ $t \in \mathbb{N}$. $\boldsymbol{d}_{i,t} \in \mathbb{R}^{n_i}$ is a known exogenous input which is periodically time-varying as well $\boldsymbol{d}_{i,t} = \boldsymbol{d}_{i,t+P}$ for all $t \in \mathbb{N}$. The set $\mathcal{N}_i$ is defined as before. The dynamics of the subsystems are coupled via the inputs.

System (3.3) can be recast as the global model (2.5) by $\boldsymbol{x} = \mathrm{col}_{i\in\mathbb{Z}_{1:M}}(\boldsymbol{x}_i)$, $\boldsymbol{u} = \mathrm{col}_{i\in\mathbb{Z}_{1:M}}(\boldsymbol{u}_i)$, $\boldsymbol{d} = \mathrm{col}_{i\in\mathbb{Z}_{1:M}}(\boldsymbol{d}_i)$, the global system matrix $\boldsymbol{A}_t = \mathrm{diag}_{i\in\mathbb{Z}_{1:M}}(\boldsymbol{A}_{ii,t})$ and the input matrix $\boldsymbol{B}_t \in \mathbb{R}^{n\times m}$ consists of the blocks $\boldsymbol{B}_{ij,t}$, i.e., $\boldsymbol{B}_t = (\boldsymbol{B}_{ij,t})$.

The states and inputs of the subsystems are subject to the decoupled possibly time-varying constraints

$$\boldsymbol{x}_{i,t} \in \mathcal{X}_{i,t} \subseteq \mathbb{R}^{n_i} \text{ and } \boldsymbol{u}_{i,t} \in \mathcal{U}_{i,t} \subseteq \mathbb{R}^{m_i} \quad \forall i \in \mathbb{Z}_{1:M},\ \forall t \in \mathbb{N}, \tag{3.4}$$

respectively, as well as to n_c coupled possibly time-varying constraints

$$\left(\mathrm{col}_{i\in\mathcal{N}_k^{\mathrm{CX}}}(\boldsymbol{x}_{i,t}), \mathrm{col}_{i\in\mathcal{N}_k^{\mathrm{CU}}}(\boldsymbol{u}_{i,t})\right) \in \mathcal{Z}_{k,t}^{\mathrm{C}} \subseteq \mathbb{R}^{n_k^{\mathrm{C}}} \quad \forall k \in \mathbb{Z}_{1:n_c},\ \forall t \in \mathbb{N} \tag{3.5}$$

where $n_k^{\mathrm{C}} = \sum_{i\in\mathcal{N}_k^{CX}} n_i + \sum_{i\in\mathcal{N}_k^{CU}} m_i$. The sets $\mathcal{N}_k^{\mathrm{CX}}$ and $\mathcal{N}_k^{\mathrm{CU}}$ contain the indexes of subsystems whose local state or input are involved in constraint k, respectively. They are assumed to be time-invariant. The sets $\mathcal{X}_{i,t}$, $\mathcal{U}_{i,t}$ and $\mathcal{Z}_{k,t}^{\mathrm{C}}$ are compact sets and assumed to be periodically time-varying, i.e., $\mathcal{X}_{i,t} = \mathcal{X}_{i,t+P}$, $\mathcal{U}_{i,t} = \mathcal{U}_{i,t+P}$ for all $i \in \mathbb{Z}_{1:M}$ and $\mathcal{Z}_{k,t}^{\mathrm{C}} = \mathcal{Z}_{k,t+P}^{\mathrm{C}}$ for all $i \in \mathbb{Z}_{1:M}$ for all $t \in \mathbb{N}$. Combining constraints (3.4) for all $i \in \mathbb{Z}_{1:M}$ and (3.5) for all $k \in \mathbb{Z}_{1:n_c}$, one can recast them as the global constraint (2.6).

3.3 Model of the Communication Network

This section discusses the main assumptions on the communication network.

Since NCSs and distributed control systems are spatially distributed, a clock synchronization protocol is required to provide a common notion of time in the components. More precisely, the common clock is required in the sensor units to have a synchronized sampling of the subsystems. Furthermore, it is needed in the receiving units of the network so that they know that no value has been sent by the transmitting system in the case of no event. Additionally, the actuation of the plant can be synchronized.

Clock synchronization can be realized, for both wired and wireless networks, by software-based or hardware-based synchronization protocols. Moreover, mixed software/hardware protocols have been developed. An overview can be found in the survey papers [SBK05, Fai07, SF11] and the references therein. Throughout the thesis the following assumption is made.

Assumption 3.3.1. *The clocks of controller, sensor and actuator units of all subsystems are synchronized.*

Although the advantages of networked-based communication are appealing, there are different types of imperfections which the network induces in the control loop. A short overview of those imperfections, which follows the presentation in [Al-16, Chapter 1.3], is listed below.

- *Time-varying transmission delays*: The time required for transmitting the data including encapsulation at the sender and decapsulation at receiver side is called the transmission delay. It can be uncertain and time-varying due to varying packet size, routing mechanism, etc.
- *Time-varying sampling intervals*: As mentioned above, clock synchronization protocols are required to provide a common notion of time in the subsystems and their components. Although the deviation among them can be sufficiently small, the sampling instants can be uncertain and time-varying. Hence, also the sampling interval which is the difference between two consecutive sampling instants is time-varying.
- *Medium access constraints*: Due to the possibly large number of components in the network, it might not be possible that all components can access the network simultaneously.
- *Packet dropouts*: A packet can be lost when transiting the network due to packet collision, network node failure or congestion.

Within this thesis the following assumption on the communication over the network is made.

Assumption 3.3.2. *The communication network is ideal in the sense that there are no transmission delays, packet dropouts, quantization errors or access constraints and the sampling interval is constant.*

A network that satisfies Assumption 3.3.2 is also called an ideal network in the thesis. Assumption 3.3.2 is relaxed in Chapter 4 where the case of bounded time-varying transmission delays and sampling intervals are considered.

Note that Assumptions 3.3.1 and 3.3.2 are made (mostly implicitly) in the vast majority of the publications on event-triggered and distributed MPC (cf. the literature review on networked MPC in Section 1.1 for centralized and Section 1.2 for distributed set-ups).

3.4 Suboptimal Robust MPC

In this section a suboptimal version of the robust MPC presented in Section 2.3 is introduced. It is of importance as in event-triggered MPC and DMPC the optimal cost function, which usually also serves as a Lyapunov function, might not be available. The reasons for that can be the event-triggered evaluation of the optimization problem or the limited available information in DMPC. As shown in the sequel, GES of an RPI set can be established despite the absence of the optimal cost function using ideas from suboptimal MPC.

The state-feedback case of (2.1) is considered. The plant is controlled by a model predictive controller exploiting the control parametrization $\boldsymbol{u}_t = \boldsymbol{v}_t + \boldsymbol{K}\boldsymbol{x}_t$. The controller is described by the mapping $\boldsymbol{h} : \mathbb{R}^n \times \mathbb{R}^{N \cdot m} \to \mathbb{R}^{N \cdot m}$. The MPC cost function is $J_N(\boldsymbol{V}_t) = \sum_{r=t}^{t+N-1} \boldsymbol{v}_r^{\mathrm{T}} \boldsymbol{M} \boldsymbol{v}_r$ where $\boldsymbol{M} \succ \boldsymbol{0}$. $\boldsymbol{V}_t \in \mathbb{R}^{m \cdot N}$ is the predicted sequence by the MPC controller.

The closed-loop system can then be recast as

$$\boldsymbol{x}_{t+1} = \boldsymbol{F}\boldsymbol{x}_t + \boldsymbol{B}\boldsymbol{v}_t + \boldsymbol{w}_t \tag{3.6a}$$

$$\boldsymbol{V}_{t+1} = \boldsymbol{h}(\boldsymbol{x}_t, \boldsymbol{V}_t) \tag{3.6b}$$

$t \in \mathbb{N}$. Let $\boldsymbol{v}_t = \boldsymbol{v}_{t|t}$ and recall that $\boldsymbol{v}_{t|t}$ is the first element of the sequence $\boldsymbol{V}_t$, the set of admissible sequences $\boldsymbol{V}$ is defined as $\mathcal{D}_N$ and the set of feasible states as $\mathcal{X}_N$. Let $\mathcal{Z} \subseteq \mathbb{R}^n$ be an RPI set for the dynamics $\boldsymbol{z}_{t+1} = \boldsymbol{F}\boldsymbol{z}_t + \boldsymbol{w}_t$. Furthermore, the function

$$V_N(\boldsymbol{x}, \boldsymbol{V}) = \min_{\boldsymbol{z} \in \mathcal{Z}} V^{\mathrm{f}}(\boldsymbol{x} - \boldsymbol{z}) + \nu J_N(\boldsymbol{V}) \tag{3.7}$$

with $V^{\mathrm{f}}(\boldsymbol{x}) = \boldsymbol{x}^{\mathrm{T}} \boldsymbol{P} \boldsymbol{x}$ and $\nu \in \mathbb{R}_{>0}$ is defined.

Now the first results can be stated.

Lemma 3.4.1. *Consider system* (3.6), *let* $\mathcal{X}_N \times \mathcal{D}_N$ *be a PI set for* (3.6) *as well as* $\boldsymbol{P}, \boldsymbol{Q} \in \mathbb{R}^{n \times n}$ *be matrices such that* $\boldsymbol{P} \succ \boldsymbol{0}$, $\boldsymbol{Q} \succ \boldsymbol{0}$ *and* $\boldsymbol{F}^{\mathrm{T}}\boldsymbol{P}\boldsymbol{F} - \boldsymbol{P} \preceq -\boldsymbol{Q}$ *holds. If*

$(\boldsymbol{x}_0, \boldsymbol{V}_0) \in \mathcal{X}_N \times \mathcal{D}_N$ and for the cost $J_N(\boldsymbol{V}_{t+1}) - J_N(\boldsymbol{V}_t) \leq -\boldsymbol{v}_t^{\mathrm{T}} \boldsymbol{M} \boldsymbol{v}_t$ holds for all $t \in \mathbb{N}$, then

$$V_N(\boldsymbol{x}_{t+1}, \boldsymbol{V}_{t+1}) - V_N(\boldsymbol{x}_t, \boldsymbol{V}_t) \leq -c_1 \|(\boldsymbol{x}_t, \boldsymbol{v}_t)\|^2_{\mathcal{Z}\times\{\boldsymbol{0}\}} \tag{3.8}$$

holds for $t \in \mathbb{N}$ where $c_1 \in \mathbb{R}_{>0}$. Moreover, $\lim_{t\to\infty} \|\boldsymbol{x}_t\|_{\mathcal{Z}} = 0$ and $\lim_{t\to\infty} \|\boldsymbol{v}_t\| = 0$.

The proof can be found in Appendix B.1.

The lemma says that the function $V_N(\boldsymbol{x}, \boldsymbol{V})$ is non-increasing along the closed-loop trajectories of (3.6) and establishes attractivity of the set $\mathcal{Z}$. Note that in general it is desired to converge to the minimal RPI set, i.e., to $\mathcal{F}^\infty$.

The next theorem establishes GES of the set $\mathcal{Z}$.

Theorem 3.4.2. *Consider system (3.6) and assume that $\mathcal{X}_N \times \mathcal{D}_N$ is a PI set for (3.6). Moreover, assume that $(\boldsymbol{x}_0, \boldsymbol{V}_0) \in \mathcal{X}_N \times \mathcal{D}_N$ and the function $V_N(\boldsymbol{x}, \boldsymbol{V})$ decreases along the closed-loop trajectory, i.e.,*

$$V_N(\boldsymbol{x}_{t+1}, \boldsymbol{V}_{t+1}) - V_N(\boldsymbol{x}_t, \boldsymbol{V}_t) \leq -c_1 \|(\boldsymbol{x}_t, \boldsymbol{v}_t)\|^2_{\mathcal{Z}\times\{\boldsymbol{0}\}} \tag{3.9}$$

for $t \in \mathbb{N}$ where $c_1 \in \mathbb{R}_{>0}$. If there exist constants $\epsilon, d \in \mathbb{R}_{>0}$ such that

$$\|\boldsymbol{V}_t\| \leq d \|\boldsymbol{x}_t\|_{\mathcal{Z}}, \forall \boldsymbol{x}_t \in \mathcal{B}_\epsilon(\boldsymbol{0}), \tag{3.10}$$

then the set $\mathcal{Z} \times \{\boldsymbol{0}\}$ is GES with ROA $\mathcal{X}_N \times \mathcal{D}_N$ for system (3.6). Moreover, the set $\mathcal{Z}$ is GES with ROA $\mathcal{X}_N$ for system (3.6a).

The proof can be found in Appendix B.1.

4 Output-based Event-triggered Model Predictive Control

In this chapter an OB-MPC strategy for decentralized control of discrete-time NCSs with event-triggered communication over the S2C and C2A channel of a network is presented. Moreover, the application to sampled-data systems with certain network imperfections is discussed. The chapter is based on [BWLG17] and [BL18b].

4.1 Problem Set-up

The NCS set-up shown in Figure 3.1 is considered. The dynamics of each subsystem $\mathscr{S}_i$ can be described by the discrete-time LTI model

$$\mathscr{S}_i : \begin{cases} \boldsymbol{x}_{i,t+1} & = \boldsymbol{A}_{ii}\boldsymbol{x}_{i,t} + \boldsymbol{B}_{ii}\boldsymbol{u}_{i,t} + \boldsymbol{w}^{\mathrm{c}}_{i,t} \\ \boldsymbol{w}^{\mathrm{c}}_{i,t} & = \sum_{j\in\mathcal{N}_i} \boldsymbol{A}_{ij}\boldsymbol{x}_{j,t} + \boldsymbol{B}_{ij}\boldsymbol{u}_{j,t} \\ \boldsymbol{y}_{i,t} & = \boldsymbol{C}_i\boldsymbol{x}_{i,t}. \end{cases} \tag{4.1}$$

Each plant can measure the output $\boldsymbol{y}_i$. $\boldsymbol{w}^{\mathrm{c}}_{i,t}$ comprises the coupling terms stemming from other subsystems. The input and state of each subsystem are subject to the constraints $\boldsymbol{u}_{i,t} \in \mathcal{U}_i$ and $\boldsymbol{x}_{i,t} \in \mathcal{X}_i$ for all $t \in \mathbb{N}$, respectively, where $\mathcal{U}_i \subseteq \mathbb{R}^{m_i}$ and $\mathcal{X}_i \subseteq \mathbb{R}^{n_i}$ are convex compact sets containing the origin in their interior. For a more detailed overview of the plant model refer to Section 3.2. Note that for reasons of simplicity, the undisturbed case is considered here. The presented approach can be extended to plants with persistent additive disturbances. Of course only convergence of the state of the closed-loop to an RPI set and not to the origin can be shown in this case.

It is assumed that all components in the NCS are synchronized as stated in Assumption 3.3.1. Moreover, at first Assumptions 3.3.2 holds which means that the communication network is ideal.

The dynamics of the overall plant $\mathscr{S}$ are given by

$$\mathscr{S} : \begin{cases} \boldsymbol{x}_{t+1} & = \boldsymbol{A}\boldsymbol{x}_t + \boldsymbol{B}\boldsymbol{u}_t \\ \boldsymbol{y}_t & = \boldsymbol{C}\boldsymbol{x}_t. \end{cases} \tag{4.2}$$

as $\tilde{\boldsymbol{x}}_{i,t} = \boldsymbol{x}_{i,t} - \hat{\boldsymbol{x}}_{i,t}$ and its dynamics are

$$\tilde{\boldsymbol{x}}_{i,t+1} = \boldsymbol{A}_{L,i}\tilde{\boldsymbol{x}}_{i,t} + \boldsymbol{w}^{\mathrm{o}}_{i,t} \tag{4.6a}$$

$$\boldsymbol{w}^{\mathrm{o}}_{i,t} = \boldsymbol{L}_i\boldsymbol{e}_{i,t} + \boldsymbol{w}^{\mathrm{c}}_{i,t}. \tag{4.6b}$$

The matrix $\boldsymbol{A}_{L,i} = \boldsymbol{A}_{ii} - \boldsymbol{L}_i\boldsymbol{C}_i$ has to satisfy the following assumption.

Assumption 4.2.1. *The matrices $\boldsymbol{A}_{L,i}$ are stable $\forall i \in \mathbb{Z}_{1:M}$.*

Due to the event-triggering rule (4.3), $\boldsymbol{e}_{i,t} \in \mathcal{E}^{\mathrm{a}}_i$ for all $t \in \mathbb{N}$ holds. The coupling $\boldsymbol{w}^{\mathrm{c}}_{i,t}$ of the neighboring subsystems is bounded, i.e., $\boldsymbol{w}^{\mathrm{c}}_{i,t} \in \mathcal{W}^{\mathrm{c}}_i = \bigoplus_{j\in\mathcal{N}_i} \boldsymbol{A}_{ij}\mathcal{X}_j \oplus \boldsymbol{B}_{ij}\mathcal{U}_j$. Note that for a centralized implementation as in [BWLG17], $\mathcal{W}^{\mathrm{c}}_i = \emptyset$ due to the absence of unmodeled coupling. Since $\mathcal{W}^{\mathrm{o}}_i = \boldsymbol{L}_i\mathcal{E}^{\mathrm{a}}_i \oplus \mathcal{W}^{\mathrm{c}}_i$ is bounded and $\boldsymbol{A}_{L,i}$ are stable matrices, there exists a non-empty compact RPI set $\Omega_i \in \mathbb{R}^{n_i}$ for (4.6) (see, e.g., [RM09]).

The observer dynamics (4.5) can be recast as

$$\hat{\boldsymbol{x}}_{i,t+1} = \boldsymbol{A}_{ii}\hat{\boldsymbol{x}}_{i,t} + \boldsymbol{B}_{ii}\boldsymbol{u}_{i,t} + \boldsymbol{w}_{i,t} \tag{4.7a}$$

$$\boldsymbol{w}_{i,t} = \boldsymbol{L}_i\boldsymbol{C}_i\tilde{\boldsymbol{x}}_{i,t} - \boldsymbol{L}_i\boldsymbol{e}_{i,t} \tag{4.7b}$$

where $\boldsymbol{w}_{i,t} \in \mathcal{W}_i = \boldsymbol{L}_i\boldsymbol{C}_i\Omega_i \oplus (-\boldsymbol{L}_i\mathcal{E}^{\mathrm{a}}_i)$. Note that $\mathcal{W}_i$ is bounded as $\mathcal{E}^{\mathrm{a}}_i$ and Ω_i are bounded. From (4.6) it follows that the actual state is in the proximity of the observer state, i.e., $\boldsymbol{x}_{i,t} \in \{\hat{\boldsymbol{x}}_{i,t}\} \oplus \Omega_i, \forall t \in \mathbb{N}$. Thus, the main idea in OB-MPC is to control the observer system (4.7) using robust MPC methods.

4.2.4 Model Predictive Controller

For each subsystem a controller $\mathscr{C}_i$ is designed for the dynamics (4.7) according to the robust MPC scheme of [CRZ01] presented in Section 2.3. At their core they solve the optimization problem

$$
\begin{aligned}
&\underset{\boldsymbol{V}_{i,t}}{\text{minimize}} && J_{N,i}(\boldsymbol{V}_{i,t}) && (4.8a)\\
&\text{subject to} && \bar{\boldsymbol{x}}_{i,t|t} = \hat{\boldsymbol{x}}_{i,t}, && (4.8b)\\
& && \bar{\boldsymbol{x}}_{i,t+N|t} \in \mathcal{X}^{\mathrm{f}}_i, && (4.8c)\\
& && \text{and } \forall r \in \mathbb{Z}_{0:N-1}: && (4.8d)\\
& && \bar{\boldsymbol{x}}_{i,t+r+1|t} = \boldsymbol{A}_{ii}\bar{\boldsymbol{x}}_{i,t+r|t} + \boldsymbol{B}_{ii}\bar{\boldsymbol{u}}_{i,t+r|t}, && (4.8e)\\
& && \bar{\boldsymbol{u}}_{i,t+r|t} = \boldsymbol{v}_{i,t+r|t} + \boldsymbol{K}_i\bar{\boldsymbol{x}}_{i,t+r|t}, && (4.8f)\\
& && \bar{\boldsymbol{x}}_{i,t+r|t} \in \bar{\mathcal{X}}^{r}_i, && (4.8g)\\
& && \bar{\boldsymbol{u}}_{i,t+r|t} \in \bar{\mathcal{U}}^{r}_i. && (4.8h)
\end{aligned}
$$

Stacking up $\boldsymbol{v}_{i,r|t}$ one obtains $\boldsymbol{V}_{i,t}^{\mathrm{T}} = \left[\boldsymbol{v}_{i,t|t}^{\mathrm{T}}, \dots, \boldsymbol{v}_{i,t+N-1|t}^{\mathrm{T}}\right]$. $\bar{\mathcal{X}}_i^r \subseteq \mathbb{R}^{n_i}$ and $\bar{\mathcal{U}}_i^r \subseteq \mathbb{R}^{m_i}$ are tightened constraint sets of the state and input constraints, respectively, and $\mathcal{X}_i^{\mathrm{f}} \subseteq \mathbb{R}^{n_i}$ is the terminal set. For more information refer to Section 2.3. $\boldsymbol{K}_i \in \mathbb{R}^{m_i \times n_i}$ is a static feedback matrix which has to satisfy the following condition.

Assumption 4.2.2. *The matrices $\boldsymbol{A}_{ii} + \boldsymbol{B}_{ii}\boldsymbol{K}_i$ are stable $\forall i \in \mathbb{Z}_{1:M}$.*

The constraint sets $\bar{\mathcal{X}}_i^r \subseteq \mathbb{R}^{n_i}$ and $\bar{\mathcal{U}}_i^r \subseteq \mathbb{R}^{m_i}$ are given by

$$\bar{\mathcal{X}}_i^r = \mathcal{X}_i \ominus \Omega_i \ominus \mathcal{F}_i^r \qquad \forall r \in \mathbb{Z}_{0:N}, \tag{4.9a}$$

$$\bar{\mathcal{U}}_i^r = \mathcal{U}_i \ominus \boldsymbol{K}_i \mathcal{F}_i^r \qquad \forall r \in \mathbb{Z}_{0:N}, \tag{4.9b}$$

$$\mathcal{F}_i^r = \bigoplus_{s=0}^{r-1} \left(\boldsymbol{A}_{ii} + \boldsymbol{B}_{ii}\boldsymbol{K}_i\right)^s \mathcal{W}_i \qquad \forall r \in \mathbb{Z}_{0:N}. \tag{4.9c}$$

The terminal sets $\mathcal{X}_i^{\mathrm{f}}$ satisfy the following assumption.

Assumption 4.2.3. *For all $i \in \mathbb{Z}_{1:M}$, there exists a non-empty set $\mathcal{X}_i^{\mathrm{f}}$ such that*

$$\boldsymbol{K}_i \mathcal{X}_i^{\mathrm{f}} \subseteq \bar{\mathcal{U}}_i^N, \quad \mathcal{X}_i^{\mathrm{f}} \subseteq \bar{\mathcal{X}}_i^N, \quad \left(\boldsymbol{A}_{ii} + \boldsymbol{B}_{ii}\boldsymbol{K}_i\right)\mathcal{X}_i^{\mathrm{f}} \oplus \left(\boldsymbol{A}_{ii} + \boldsymbol{B}_{ii}\boldsymbol{K}_i\right)^N \mathcal{W}_i \subseteq \mathcal{X}_i^{\mathrm{f}}. \tag{4.10}$$

The assumption is identical to Assumption 2.3.2 in the centralized case.

In addition, the following definitions, which are borrowed from the centralized case, are needed. The set of admissible sequences $\boldsymbol{V}_i$ is defined as

$$\mathcal{D}_{N,i}\left(\boldsymbol{x}_i\right) = \left\{\boldsymbol{V}_i \in \mathbb{R}^{m_i \cdot N} | (4.8) \text{ is feasible}\right\}. \tag{4.11}$$

Furthermore, the set of states for which a feasible sequence $\boldsymbol{V}_i$ can be found such that all constraints are feasible is defined as

$$\mathcal{X}_{N,i} = \left\{\boldsymbol{x}_i \in \mathbb{R}^{n_i} | \mathcal{D}_{N,i}\left(\boldsymbol{x}_i\right) \neq \emptyset\right\}. \tag{4.12}$$

Another important set is

$$\begin{aligned}\mathcal{X}_{K,i} = \Big\{\boldsymbol{x}_i \in \mathbb{R}^{n_i} | \left(\boldsymbol{A}_{ii} + \boldsymbol{B}_{ii}\boldsymbol{K}_i\right)^N \boldsymbol{x}_i \in \mathcal{X}_i^{\mathrm{f}}, \left(\boldsymbol{A}_{ii} + \boldsymbol{B}_{ii}\boldsymbol{K}_i\right)^r \boldsymbol{x}_i \in \bar{\mathcal{X}}_i^r,\\ \boldsymbol{K}_i\left(\boldsymbol{A}_{ii} + \boldsymbol{B}_{ii}\boldsymbol{K}_i\right)^r \boldsymbol{x}_i \in \bar{\mathcal{U}}_i^r \ \forall r \in \mathbb{Z}_{0:N-1}\Big\}\end{aligned} \tag{4.13}$$

which is the set of states where $\boldsymbol{v}_{i,t+r|t} = \boldsymbol{0}$ for $r \in \mathbb{Z}_{0:N-1}$ is a feasible solution to (4.8). Obviously, $\mathcal{X}_{K,i} \subseteq \mathcal{X}_{N,i}$ holds for all $i \in \mathbb{Z}_{1:M}$.

The cost function of optimization problem (4.8) is

$$J_{N,i}\left(\boldsymbol{V}_{i,t}\right) = \sum_{r=t}^{t+N-1} \boldsymbol{v}_{i,r|t}^{\mathrm{T}} \boldsymbol{M}_i \boldsymbol{v}_{i,r|t} \tag{4.14}$$

where $\boldsymbol{M}_i \in \mathbb{R}^{m_i \times m_i}$ with $\boldsymbol{M}_i \succ \mathbf{0}$ is a weighting matrix. $J^*_{N,i}(\hat{\boldsymbol{x}}_{i,t})$ denotes the optimal cost function, i.e.,

$$J^*_{N,i}(\hat{\boldsymbol{x}}_{i,t}) = \min_{\boldsymbol{V}_{i,t} \in \mathcal{D}_{N,i}(\hat{\boldsymbol{x}}_{i,t})} J_{N,i}(\boldsymbol{V}_{i,t}). \tag{4.15}$$

The optimal sequences $\boldsymbol{V}^*_{i,t}(\hat{\boldsymbol{x}}_{i,t}) = \arg\min_{\boldsymbol{V}_{i,t} \in \mathcal{D}_{N,i}(\hat{\boldsymbol{x}}_{i,t})} J_{N,i}(\boldsymbol{V}_{i,t})$. The optimal input sequence and state sequence $\bar{\boldsymbol{U}}^*_{i,t}$ and $\bar{\boldsymbol{X}}^*_{i,t}$ can be constructed from that.

The actuator $\mathscr{A}_i$ applies the first input of the optimal input sequence $\bar{\boldsymbol{U}}^*_{i,t}$ to the plant:

$$\mathscr{A}_i : \boldsymbol{u}_{i,t} = \bar{\boldsymbol{u}}^*_{i,t|t}(\hat{\boldsymbol{x}}_{i,t}). \tag{4.16}$$

Next the following centralized representatives are introduced for the further analysis: $\hat{\boldsymbol{x}} = \operatorname{col}_{i \in \mathbb{Z}_{1:M}}(\hat{\boldsymbol{x}}_i)$, $\tilde{\boldsymbol{x}} = \operatorname{col}_{i \in \mathbb{Z}_{1:M}}(\tilde{\boldsymbol{x}}_i)$, $\boldsymbol{w} = \operatorname{col}_{i \in \mathbb{Z}_{1:M}}(\boldsymbol{w}_i)$, $\mathcal{X}^{\mathrm{f}} = \prod_{i=1}^M \mathcal{X}^{\mathrm{f}}_i$, $\mathcal{X}_N = \prod_{i=1}^M \mathcal{X}_{N,i}$, $\bar{\mathcal{X}}^l = \prod_{i=1}^M \bar{\mathcal{X}}^l_i$, $\bar{\mathcal{U}}^l = \prod_{i=1}^M \bar{\mathcal{U}}^l_i$ for $l \in \mathbb{Z}_{0:N}$, $\mathcal{X}_K = \prod_{i=1}^M \mathcal{X}_{K,i}$, $\Omega = \prod_{i=1}^M \Omega_i$, $\mathcal{W}^{\mathrm{o}} = \prod_{i=1}^M \mathcal{W}^{\mathrm{o}}_i$, $\boldsymbol{K} = \operatorname{diag}_{i \in \mathbb{Z}_{1:M}}(\boldsymbol{K}_i)$ and $J^*_N(\hat{\boldsymbol{x}}_t) = \sum_{i=1}^M \min_{\boldsymbol{V}_{i,t} \in \mathcal{D}_{N,i}(\hat{\boldsymbol{x}}_{i,t})} J_{N,i}(\boldsymbol{V}_{i,t})$.

Remark 4.2.4. *On the one hand a larger absolute trigger set $\mathcal{E}^{\mathrm{a}}_i$ is expected to lead to less trigger events and thus reduces the communicational burden. On the other hand the sizes of $\mathcal{W}_i$ and Ω_i increase which decreases the ROA of the controller. Obviously, a trade-off has to be found when selecting $\mathcal{E}^{\mathrm{a}}_i$.*

Moreover, the sets Ω_i and $\mathcal{W}_i$ depend on the coupling matrices $\boldsymbol{A}_{ij}$ and $\boldsymbol{B}_{ij}$. Loosely speaking, their norms have to be small (i.e., weak coupling between the subsystems) so that the size of Ω_i and $\mathcal{W}_i$ are small and Assumption 4.2.3 can be satisfied. Note that this is a standard assumption in DeMPC.

4.3 Analysis MPC Strategy for S2C Channel

In this section the main properties of the proposed method are discussed. First, the results concerning recursive feasibility and constraint satisfaction are stated.

Theorem 4.3.1. *Assume $\hat{\boldsymbol{x}}_0 \in \mathcal{X}_N$ as well as $\tilde{\boldsymbol{x}}_0 \in \Omega$ and let the output measurements $\boldsymbol{y}^{\mathrm{ET}}_{i,t}$ be updated according to (4.3) for all $i \in \mathbb{Z}_{1:M}$ and $t \in \mathbb{N}$, then $\hat{\boldsymbol{x}}_t \in \mathcal{X}_N$, $\hat{\boldsymbol{x}}_t \in \mathcal{X} \ominus \Omega$, $\boldsymbol{u}_t \in \mathcal{U}$ and $\boldsymbol{x}_t \in \mathcal{X}$ for all $t \in \mathbb{N}$.*

The proof can be found in Appendix B.2.

Collecting dynamics (4.7) for all subsystems, one obtains

$$\hat{\boldsymbol{x}}_{t+1} = \boldsymbol{A}_{\mathrm{D}} \hat{\boldsymbol{x}}_t + \boldsymbol{B}_{\mathrm{D}} \boldsymbol{u}_t + \boldsymbol{w}_t \tag{4.17}$$

where $\boldsymbol{A}_{\mathrm{D}} = \mathrm{diag}_{i\in\mathbb{Z}_{1:M}}(\boldsymbol{A}_{ii})$ and $\boldsymbol{B}_{\mathrm{D}} = \mathrm{diag}_{i\in\mathbb{Z}_{1:M}}(\boldsymbol{B}_{ii})$. Collecting dynamics (4.1), (4.5) and (4.16) for all subsystems $i \in \mathbb{Z}_{1:M}$, the following augmented model is obtained:

$$\boldsymbol{\phi}_{t+1} = \check{\boldsymbol{A}}\boldsymbol{\phi}_t + \check{\boldsymbol{B}}\boldsymbol{u}_t + \check{\boldsymbol{L}}\boldsymbol{e}_t \tag{4.18}$$

$$\text{where } \boldsymbol{\phi}_t = \begin{pmatrix}\tilde{\boldsymbol{x}}_t\\ \hat{\boldsymbol{x}}_t\end{pmatrix},\ \check{\boldsymbol{A}} = \begin{pmatrix}\boldsymbol{A}_{\mathrm{L}} & \boldsymbol{A}_{\mathrm{C}}\\ \boldsymbol{L}\boldsymbol{C} & \boldsymbol{A}_{\mathrm{D}}\end{pmatrix},\ \check{\boldsymbol{B}} = \begin{pmatrix}\boldsymbol{B}_{\mathrm{C}}\\ \boldsymbol{B}_{\mathrm{D}}\end{pmatrix},\ \check{\boldsymbol{L}} = \begin{pmatrix}\boldsymbol{L}\\ -\boldsymbol{L}\end{pmatrix}$$

and $\boldsymbol{A}_{\mathrm{L}} = \boldsymbol{A} - \boldsymbol{L}\boldsymbol{C}$, $\boldsymbol{B}_{\mathrm{C}} = \boldsymbol{B} - \boldsymbol{B}_{\mathrm{D}}$, $\boldsymbol{A}_{\mathrm{C}} = \boldsymbol{A} - \boldsymbol{A}_{\mathrm{D}}$, $\boldsymbol{L} = \mathrm{diag}_{i\in\mathbb{Z}_{1:M}}(\boldsymbol{L}_i)$. The input applied to (4.18) is

$$\boldsymbol{u}_t = \mathrm{col}_{i\in\mathbb{Z}_{1:M}}\left(\bar{\boldsymbol{u}}^*_{i,t|t}(\hat{\boldsymbol{x}}_{i,t})\right). \tag{4.19}$$

Due to the relative part of the event triggers $\mathscr{E}\mathscr{T}_{\mathrm{s},i}$,

$$\boldsymbol{e}_t^{\mathrm{T}}\boldsymbol{S}^{\mathrm{s}}\boldsymbol{e}_t \leq \boldsymbol{\phi}_t^{\mathrm{T}}\check{\boldsymbol{S}}^{\mathrm{s}}\boldsymbol{\phi}_t \quad \forall t \in \mathbb{N} \tag{4.20}$$

holds where $\check{\boldsymbol{S}}^{\mathrm{s}} = \sigma^{\mathrm{s}}\check{\boldsymbol{C}}^{\mathrm{T}}\boldsymbol{S}^{\mathrm{s}}\check{\boldsymbol{C}}$, $\check{\boldsymbol{C}} = \begin{pmatrix}\boldsymbol{C} & \boldsymbol{C}\end{pmatrix}$, $\boldsymbol{e} = \mathrm{col}_{i\in\mathbb{Z}_{1:M}}(\boldsymbol{e}_i)$ and $\boldsymbol{S}^{\mathrm{s}} = \mathrm{diag}(\boldsymbol{S}^{\mathrm{s}}_i)$. Next, some lemmas are introduced which are required for the stability investigations. For this purpose $\boldsymbol{v}_t = \mathrm{col}_{i\in\mathbb{Z}_{1:M}}\left(\bar{\boldsymbol{v}}^*_{i,t|t}(\hat{\boldsymbol{x}}_{i,t})\right)$ is defined.

Lemma 4.3.2. *Consider the closed-loop system* (4.17) *and* (4.19) *and assume that* $\hat{\boldsymbol{x}}_0 \in \mathcal{X}_N$, *then for the optimal cost*

$$J_N^*(\hat{\boldsymbol{x}}_{t+1}) - J_N^*(\hat{\boldsymbol{x}}_t) \leq -\boldsymbol{v}_t^{\mathrm{T}}\boldsymbol{M}\boldsymbol{v}_t \tag{4.21}$$

$\forall t \in \mathbb{N}$ *holds.*

The statement follows from applying Lemma 2.3.3.

Furthermore, the sets $\Phi_K = \Omega \times \mathcal{X}_K$ and $\Phi_N = \Omega \times \mathcal{X}_N$ are defined.

Lemma 4.3.3. *The following statements hold:*

(a) The sets $\mathcal{X}_K$ and Φ_K are non-empty.

(b) The set $\mathcal{X}_K$ is an RPI set for the dynamics (4.17) *and* (4.19).

(c) The set Φ_K is an RPI set for the dynamics (4.18) *and* (4.19).

The proof can be found in Appendix B.2.

Moreover, the function $V_N^*(\boldsymbol{\phi}_t) = V^{\mathrm{f}}(\boldsymbol{\phi}_t) + \nu J_N^*(\hat{\boldsymbol{x}}_t)$ is defined where $\nu > 0$, $V^{\mathrm{f}}(\boldsymbol{\phi}_t) = \boldsymbol{\phi}_t^{\mathrm{T}}\check{\boldsymbol{P}}\boldsymbol{\phi}_t$ with $\check{\boldsymbol{P}} \in \mathbb{R}^{2n\times 2n}$. For establishing attractivity of the origin, the following main assumption is required.

Assumption 4.3.4. *Assume that the following matrix inequality is feasible*

$$\begin{pmatrix} \boldsymbol{\Psi}_{11} & \boldsymbol{\Psi}_{12} \\ * & \boldsymbol{\Psi}_{22} \end{pmatrix} \succeq \mathbf{0} \tag{4.22a}$$

$$\boldsymbol{\Psi}_{11} = \check{\boldsymbol{P}} - \check{\boldsymbol{A}}_K^{\mathrm{T}} \check{\boldsymbol{P}} \check{\boldsymbol{A}}_K - \sigma^{\mathrm{s}} \check{\boldsymbol{C}}^{\mathrm{T}} \boldsymbol{S}^{\mathrm{s}} \check{\boldsymbol{C}} - \delta \check{\boldsymbol{Q}} \tag{4.22b}$$

$$\boldsymbol{\Psi}_{12} = -\check{\boldsymbol{A}}_K^{\mathrm{T}} \check{\boldsymbol{P}} \check{\boldsymbol{L}} \tag{4.22c}$$

$$\boldsymbol{\Psi}_{22} = \boldsymbol{S}^{\mathrm{s}} - \check{\boldsymbol{L}}^{\mathrm{T}} \check{\boldsymbol{P}} \check{\boldsymbol{L}} - \check{\boldsymbol{N}} \tag{4.22d}$$

with $\check{\boldsymbol{P}} \succ \mathbf{0}$, $\check{\boldsymbol{Q}} \succ \mathbf{0}$, $\check{\boldsymbol{N}} \succ \mathbf{0}$, $0 < \delta < 1$, $\check{\boldsymbol{A}}_K = \check{\boldsymbol{A}} + \check{\boldsymbol{B}} \check{\boldsymbol{K}}$ *and* $\check{\boldsymbol{K}} = [\mathbf{0}\ \boldsymbol{K}]$.

Next an interpretation of Assumption 4.3.4 and ways how to exploit it for the design of the event triggers are discussed. The next lemma shows that Assumption 4.3.4 assures that the function $V^{\mathrm{f}}(\boldsymbol{\phi})$ is non-increasing along closed-loop trajectories starting from the set Φ_K.

Lemma 4.3.5. *Assume* $\boldsymbol{\phi}_0 \in \Phi_K$ *and let the output measurements* $\boldsymbol{y}_{i,t}^{\mathrm{ET}}$ *be updated according to* (4.3) $\forall i \in \mathbb{Z}_{1:M}$ *and* $t \in \mathbb{N}$. *Then along the closed-loop trajectory of system* (4.18) *and* (4.19)

$$V^{\mathrm{f}}(\boldsymbol{\phi}_{t+1}) - V^{\mathrm{f}}(\boldsymbol{\phi}_t) \leq -\delta \boldsymbol{\phi}_t^{\mathrm{T}} \check{\boldsymbol{Q}} \boldsymbol{\phi}_t - \boldsymbol{e}_t^{\mathrm{T}} \check{\boldsymbol{N}} \boldsymbol{e}_t \tag{4.23}$$

$\forall t \in \mathbb{N}$ *holds.*

The proof can be found in Appendix B.2. To obtain suitable trigger parameters $\boldsymbol{S}^{\mathrm{s}}$ and σ^{s} which fulfill the matrix inequality (4.22) in Assumption 4.3.4, an Linear Matrix Inequality (LMI) optimization problem can be solved. Maximizing σ^{s} in the objective function, can lead to a reduced number of transmissions triggered by the relative trigger. Note that the solution of the optimization problem is dependent on δ. Obviously, the bigger δ (faster decay of the function $V^{\mathrm{f}}(\boldsymbol{\phi})$, see (4.23)), the smaller is σ^{s} (higher communication effort due to the relative trigger).

The next lemma provides a sufficient condition for fulfilling Assumption 4.3.4.

Lemma 4.3.6. *Suppose that* $\check{\boldsymbol{A}}_K$ *is stable, then there exist* $\boldsymbol{S}^{\mathrm{s}} \succ \mathbf{0}$ *and* $\sigma^{\mathrm{s}} > 0$ *such that* (4.22) *is feasible.*

The proof can again be found in Appendix B.2. Note that Assumptions 4.2.1 and 4.2.2 only imply that $\check{\boldsymbol{A}}_K$ is stable in the centralized case, i.e., $M = 1$. If $M > 1$ the condition $\check{\boldsymbol{A}}_K$ being stable is more likely to be met when the subsystems are weakly coupled.

Next the stability property is stated.

Theorem 4.3.7. *Assume* $\hat{\boldsymbol{x}}_0 \in \mathcal{X}_N$ *as well as* $\tilde{\boldsymbol{x}}_0 \in \Omega$ *and let* $\boldsymbol{y}_{i,t}^{\mathrm{ET}}$ *be updated according to* (4.3) $\forall i \in \mathbb{Z}_{1:M}$ *and* $t \in \mathbb{N}$, *then* $V_N^*(\boldsymbol{\phi})$ *is a Lyapunov function for the closed-loop system* (4.18) *and* (4.19) *and the origin is GES with ROA* Φ_N.

The proof can be found in Appendix B.2.

4.4 MPC Strategy for S2C and C2A Channel

4.4.1 Problem Set-up

The networked configuration in Figure 4.2 is considered where communication over the C2A channel is included again. The strategy presented in Section 4.2 is now extended with an event trigger where new input sequences $\bar{\boldsymbol{U}}_{t_{i,r}}^*$ are only recomputed and sent over the network to the buffered actuators $\mathscr{B}_{\mathrm{a},i}$ located in the actuator units when it is required. $t_{i,r}, r \in \mathbb{N}$ is the time at which the last event has been triggered in the event trigger $\mathscr{E}\mathscr{T}_{\mathrm{c},i}$. $\mathscr{E}\mathscr{T}_{\mathrm{c},i}$ is located in the controller unit and monitors the deviation between the predicted and actual state of the observer. For this purpose, the predicted state sequence $\bar{\boldsymbol{X}}_{i,t_{i,r}}^*$ has to be buffered in the buffer $\mathscr{B}_{\mathrm{c},i}$ which is also placed in the controller unit. Moreover, the predicted input sequence $\bar{\boldsymbol{U}}_{t_{i,r}}^*$ has to be buffered there. The buffers $\mathscr{B}_{\mathrm{c},i}$ and $\mathscr{B}_{\mathrm{a},i}$ apply the same input to the observer as to the plant.

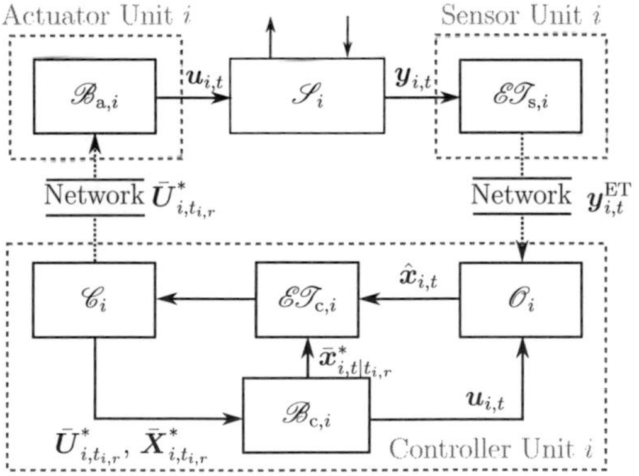

Figure 4.2: Networked control configuration with S2C and C2A channel

The triggering rule is

$$\mathscr{E}\mathscr{T}_{\mathrm{c},i}: \left\{ \begin{array}{l} \bar{\boldsymbol{U}}^*_{i,t_{i,r}} \text{ is computed and sent over the C2A channel} \\ \Leftrightarrow \hat{\boldsymbol{x}}_{i,t} - \bar{\boldsymbol{x}}^*_{i,t|t_{i,r}} \notin \mathcal{G}_i^{t-t_{i,r}}\left(\hat{\boldsymbol{x}}_{i,t}\right) \end{array} \right. \tag{4.24}$$

in every subsystem $\mathscr{S}_i$ with $i \in \mathbb{Z}_{1:M}$ and $t \in \mathbb{Z}_{t_{i,r}:t_{i,r}+N}$. The new input sequence is computed by solving the optimization problem (4.8). The trigger set $\mathcal{G}_i^r\left(\hat{\boldsymbol{x}}\right)$ is again the intersection of a relative trigger and an absolute set. The absolute sets $\mathcal{G}_i^{\mathrm{a},r}$ can be chosen such that

$$\boldsymbol{A}_{ii}\mathcal{G}_i^{\mathrm{a},r} \oplus \mathcal{W}_i \subseteq \mathcal{F}_i^{r+1}, \quad \forall r \in \mathbb{Z}_{0:N-1} \tag{4.25}$$

holds for $i \in \mathbb{Z}_{1:M}$ where $\mathcal{G}_i^{\mathrm{a},0} = \{\boldsymbol{0}\}$ and $\mathcal{G}_i^{\mathrm{a},N} = \emptyset$. The relative sets are $\mathcal{G}_i^{\mathrm{r}}\left(\hat{\boldsymbol{x}}_i\right) = \left\{\boldsymbol{x}_i \in \mathbb{R}^{n_i} | \boldsymbol{x}_i^{\mathrm{T}}\boldsymbol{S}_i^{\mathrm{c}}\boldsymbol{x} \leq \sigma^{\mathrm{c}}\hat{\boldsymbol{x}}_i^{\mathrm{T}}\boldsymbol{S}_i^{\mathrm{c}}\hat{\boldsymbol{x}}_i\right\}$ where $\sigma^{\mathrm{c}} \in \mathbb{R}_{\geq 0}$ is again a parameter for adjusting the communication effort and $\boldsymbol{S}_i^{\mathrm{c}} \in \mathbb{R}^{n_i \times n_i}$ with $\boldsymbol{S}_i^{\mathrm{c}} \succ \boldsymbol{0}$ are scaling matrices. The input applied to the plant is

$$\begin{aligned} \boldsymbol{u}_t &= \mathrm{col}_{i\in\mathbb{Z}_{1:M}}\left(\bar{\boldsymbol{u}}^*_{i,t|t_{i,r}}\left(\hat{\boldsymbol{x}}_{i,t_{i,r}}\right)\right) \\ t_{i,r+1} &= \min\left\{t \in \mathbb{N} | t > t_{i,r} \wedge \hat{\boldsymbol{x}}_{i,t} - \bar{\boldsymbol{x}}^*_{i,t|t_{i,r}} \notin \mathcal{G}_i^{t-t_{i,r}}\left(\hat{\boldsymbol{x}}_{i,t}\right)\right\} \end{aligned} \tag{4.26}$$

where $t_{i,0} = 0$ for all $i \in \mathbb{Z}_{1:M}$. Note that for the absolute trigger sets ideas from [BHA17] are exploited which can be seen as an event-triggered implementation of the robust MPC method of [CRZ01]. Moreover, it is worth mentioning that through the event-triggering the computational load is also reduced as the optimization problem does not have to be solved every time step.

4.4.2 Analysis

The relative event trigger set ensures that the deviation between predicted and actual state, i.e., $\boldsymbol{g}_t = \hat{\boldsymbol{x}}_t - \bar{\boldsymbol{x}}_t$ with $\bar{\boldsymbol{x}}_t = \mathrm{col}_{i\in\mathbb{Z}_{1:M}}\left(\bar{\boldsymbol{x}}^*_{i,t|t_{i,r}}\right)$ is bounded by $\boldsymbol{g}_t^{\mathrm{T}}\boldsymbol{S}^{\mathrm{c}}\boldsymbol{g}_t \leq \boldsymbol{\phi}_t^{\mathrm{T}}\check{\boldsymbol{S}}^{\mathrm{c}}\boldsymbol{\phi}_t$ where $\check{\boldsymbol{S}}^{\mathrm{c}} = \sigma^{\mathrm{c}}\check{\boldsymbol{T}}^{\mathrm{T}}\boldsymbol{S}^{\mathrm{c}}\check{\boldsymbol{T}}$ with $\check{\boldsymbol{T}} = \begin{pmatrix}\boldsymbol{0} & \boldsymbol{I}\end{pmatrix}$ and $\boldsymbol{S}^{\mathrm{c}} = \mathrm{diag}_{i\in\mathbb{Z}_{1:M}}\left(\boldsymbol{S}_i^{\mathrm{c}}\right)$. For establishing attractivity of the origin, the following assumption is required.

Assumption 4.4.1. *For $\check{\boldsymbol{P}} \succ \boldsymbol{0}$, $\check{\boldsymbol{Q}} \succ \boldsymbol{0}$, $\check{\boldsymbol{N}} \succ \boldsymbol{0}$, $0 < \delta < 1$, $\check{\boldsymbol{D}} = \begin{pmatrix}\check{\boldsymbol{L}} & -\check{\boldsymbol{B}}\boldsymbol{K}\end{pmatrix}$ and $\check{\boldsymbol{S}} = \mathrm{diag}\left(\boldsymbol{S}^{\mathrm{s}}, \boldsymbol{S}^{\mathrm{c}}\right)$, the following matrix inequality is feasible*

$$\begin{pmatrix}\boldsymbol{\Psi}_{11} & \boldsymbol{\Psi}_{12} \\ * & \boldsymbol{\Psi}_{22}\end{pmatrix} \succeq \boldsymbol{0} \tag{4.27a}$$

$$\boldsymbol{\Psi}_{11} = \check{\boldsymbol{P}} - \check{\boldsymbol{A}}_K^{\mathrm{T}}\check{\boldsymbol{P}}\check{\boldsymbol{A}}_K - \sigma^{\mathrm{s}}\check{\boldsymbol{C}}^{\mathrm{T}}\boldsymbol{S}^{\mathrm{s}}\check{\boldsymbol{C}} - \sigma^{\mathrm{c}}\check{\boldsymbol{T}}^{\mathrm{T}}\boldsymbol{S}^{\mathrm{c}}\check{\boldsymbol{T}} - \delta\check{\boldsymbol{Q}} \tag{4.27b}$$

$$\boldsymbol{\Psi}_{12} = \check{\boldsymbol{A}}_K^{\mathrm{T}}\check{\boldsymbol{P}}\check{\boldsymbol{D}} \tag{4.27c}$$

$$\boldsymbol{\Psi}_{22} = \check{\boldsymbol{S}} - \check{\boldsymbol{D}}^{\mathrm{T}}\check{\boldsymbol{P}}\check{\boldsymbol{D}} - \check{\boldsymbol{N}}. \tag{4.27d}$$

The interpretation of assumption is the equivalent to Assumption 4.3.4 in the S2C channel case. Similarly, an LMI optimization problem can be solved to obtain $\boldsymbol{S}^{\mathrm{s}}$, $\boldsymbol{S}^{\mathrm{c}}$, σ^{s} and σ^{c}. If $\breve{\boldsymbol{A}}_K$ is stable, then $\sigma^{\mathrm{s}} > 0$ and $\sigma^{\mathrm{c}} > 0$ can always be found. Next, the main properties of the extended approach are summarized in two theorems.

Theorem 4.4.2. *Consider the closed-loop system* (4.18) *and* (4.26). *Assume* $\hat{\boldsymbol{x}}_0 \in \mathcal{X}_N$ *as well as* $\tilde{\boldsymbol{x}}_0 \in \Omega$. *Let* $\boldsymbol{y}_{i,t}^{\mathrm{ET}}$ *and the input sequence* $\bar{\boldsymbol{U}}_i^*$ *be updated according to* (4.3) *and* (4.24) $\forall i \in \mathbb{Z}_{1:M}$ *and* $t \in \mathbb{N}$, *respectively. Then* $\hat{\boldsymbol{x}}_t \in \mathcal{X}_N$, $\hat{\boldsymbol{x}}_t \in \mathcal{X} \ominus \Omega$, $\boldsymbol{u}_t \in \mathcal{U}$ *and* $\boldsymbol{x}_t \in \mathcal{X}$ *for all* $t \in \mathbb{N}$.

Theorem 4.4.3. *Assume* $\hat{\boldsymbol{x}}_0 \in \mathcal{X}_N$ *as well as* $\tilde{\boldsymbol{x}}_0 \in \Omega$. *Let* $\boldsymbol{y}_{i,t}^{\mathrm{ET}}$ *and the input sequence* $\bar{\boldsymbol{U}}_i^*$ *be updated according to* (4.3) *and* (4.24) $\forall i \in \mathbb{Z}_{1:M}$ *and* $t \in \mathbb{N}$, *respectively. Then the origin of the closed-loop system* (4.18) *and* (4.26) *is GES with ROA* Φ_N.

The proofs can be found in Appendix B.2.

4.5 Sampled-data Systems and Network Effects

In this section, it is discussed how the presented approach can be extended to the control of sampled-data systems. Additionally, it is considered how the network-induced effects of bounded time-varying transmission delays and sampling intervals can be handled. A continuous-time LTI plant which can be described by

$$\dot{\boldsymbol{x}}(\theta) = \boldsymbol{A}_{\mathrm{c}} \boldsymbol{x}(\theta) + \boldsymbol{B}_{\mathrm{c}} \boldsymbol{u}(\theta) \tag{4.28a}$$

$$\boldsymbol{y}(\theta) = \boldsymbol{C} \boldsymbol{x}(\theta) \tag{4.28b}$$

with time $\theta \in \mathbb{R}_{\geq 0}$ and system matrix $\boldsymbol{A}_{\mathrm{c}}$ and input matrix $\boldsymbol{B}_{\mathrm{c}}$ is considered. Recall that the solution of the differential equation (4.28a) is given by

$$\boldsymbol{x}(\theta) = e^{\boldsymbol{A}_{\mathrm{c}}(\theta-\theta_t)} \boldsymbol{x}_t + \int_{\theta_t}^{\theta} e^{\boldsymbol{A}_{\mathrm{c}}(\theta-s)} \boldsymbol{B}_{\mathrm{c}} \boldsymbol{u}(s)\, ds \tag{4.29}$$

where $\boldsymbol{x}_t = \boldsymbol{x}(\theta_t)$. The output of the plant is sampled with a sampling interval $h_t = \theta_{t+1} - \theta_t$ where θ_t is the sampling instant. The sampling interval can be time-varying but is bounded, i.e., $h_t \in [h^{\min}, h^{\max}]$ with $0 < h^{\min} \leq h^{\max}$. The controller units $\mathscr{CU}_i$ work in discrete-time as they are implemented on digital platforms. The discrete-time input $\boldsymbol{u}_t$ is transfered to continuous-time in the actuator units $\mathscr{AU}_i$ using a Zero-Order Hold (ZOH) unit, i.e.,

$$\boldsymbol{u}(\theta) = \boldsymbol{u}_t \quad \theta \in [\theta_t, \theta_{t+1}). \tag{4.30}$$

Discretizing the continuous-time plant (4.28) using exact discretization, one obtains

$$\boldsymbol{x}_{t+1} = \boldsymbol{A}_{\mathrm{d}}(h_t)\,\boldsymbol{x}_t + \boldsymbol{B}_{\mathrm{d}}(h_t)\,\boldsymbol{u}_t \tag{4.31a}$$
$$\boldsymbol{y}_t = \boldsymbol{C}\boldsymbol{x}_t \tag{4.31b}$$

where $\boldsymbol{y}_t = \boldsymbol{y}(\theta_t)$, $\boldsymbol{A}_{\mathrm{d}}(h_t) = e^{\boldsymbol{A}_{\mathrm{c}} h_t}$ and $\boldsymbol{B}_{\mathrm{d}}(h_t) = \int_0^{h_t} e^{\boldsymbol{A}_{\mathrm{c}} s} ds \boldsymbol{B}_{\mathrm{c}}$ (cf. (4.29)).

System (4.31) can be considered as a system with parametric uncertainty. As the dependency on the uncertainty is nonlinear, it is reasonable to approximate it in order to obtain a formulation which is more suitable for analysis. For this purpose, the matrices $\boldsymbol{A}_{\mathrm{d}}^{[k]}$ and $\boldsymbol{B}_{\mathrm{d}}^{[k]}$ with $k \in \mathbb{Z}_{1:n_{\mathrm{pa}}}$ are defined such that

$$\begin{aligned} \boldsymbol{A}_{\mathrm{d}}(h_t) &\in \mathrm{co}\left(\left\{\boldsymbol{A}_{\mathrm{d}}^{[k]}\right\}, k \in \mathbb{Z}_{1:n_{\mathrm{pa}}}\right), \\ \boldsymbol{B}_{\mathrm{d}}(h_t) &\in \mathrm{co}\left(\left\{\boldsymbol{B}_{\mathrm{d}}^{[k]}\right\}, k \in \mathbb{Z}_{1:n_{\mathrm{pa}}}\right) \quad \forall\, h_t \in [h^{\min}, h^{\max}]. \end{aligned} \tag{4.32}$$

This means that the matrices $\boldsymbol{A}_{\mathrm{d}}(h_t)$ and $\boldsymbol{B}_{\mathrm{d}}(h_t)$ are over-approximated by convex sets. n_{pa} is the number of vertexes of the polytopic over-approximation. An over-approximation of this form can be obtained using the real Jordan form or the Cayley-Hamilton theorem, see [HVG+10] for additional information.

In the controller unit the observer

$$\mathscr{O}_i:\ \hat{\boldsymbol{x}}_{i,t+1} = \boldsymbol{A}_{\mathrm{d}ii}\left(h^{\mathrm{f}}\right)\hat{\boldsymbol{x}}_{i,t} + \boldsymbol{B}_{\mathrm{d}ii}\left(h^{\mathrm{f}}\right)\boldsymbol{u}_{i,t} + \boldsymbol{L}_i\left(\boldsymbol{y}_{i,t}^{\mathrm{ET}} - \boldsymbol{C}_i\hat{\boldsymbol{x}}_{i,t}\right). \tag{4.33}$$

is applied. The matrices $\boldsymbol{A}_{\mathrm{d}ii}\left(h^{\mathrm{f}}\right)$ and $\boldsymbol{B}_{\mathrm{d}ii}\left(h^{\mathrm{f}}\right)$ are the block-diagonal components of the matrices $\boldsymbol{A}_{\mathrm{d}}\left(h^{\mathrm{f}}\right)$ and $\boldsymbol{B}_{\mathrm{d}}\left(h^{\mathrm{f}}\right)$, respectively. $h^{\mathrm{f}} \in [h^{\min}, h^{\max}]$ is a fixed value for the sampling interval which is chosen by the designer. The observer error evolves according to

$$\tilde{\boldsymbol{x}}_{i,t+1} = \left(\boldsymbol{A}_{\mathrm{d}ii}\left(h^{\mathrm{f}}\right) - \boldsymbol{L}_i\boldsymbol{C}_i\right)\tilde{\boldsymbol{x}}_{i,t} + \boldsymbol{w}_{i,t}^{\mathrm{o}} \tag{4.34}$$
$$\boldsymbol{w}_{i,t}^{\mathrm{o}} = \boldsymbol{w}_{i,t}^{\mathrm{c}} + \boldsymbol{L}_i\boldsymbol{e}_{i,t} + \left(\boldsymbol{A}_{\mathrm{d}ii}(h_t) - \boldsymbol{A}_{\mathrm{d}ii}\left(h^{\mathrm{f}}\right)\right)\boldsymbol{x}_{i,t} + \left(\boldsymbol{B}_{\mathrm{d}ii}(h_t) - \boldsymbol{B}_{\mathrm{d}ii}\left(h^{\mathrm{f}}\right)\right)\boldsymbol{u}_{i,t} \tag{4.35}$$
$$\boldsymbol{w}_{i,t}^{\mathrm{c}} = \sum_{j\in\mathcal{N}_j} \boldsymbol{A}_{\mathrm{d}ij}(h_t)\,\boldsymbol{x}_{j,t} + \boldsymbol{B}_{\mathrm{d}ij}(h_t)\,\boldsymbol{u}_{j,t}. \tag{4.36}$$

Obviously, there is an additional error between the prediction and the actual evolution of the plant caused by the usage of the fixed value h^{f}. Nevertheless, $\boldsymbol{w}_{i,t}^{\mathrm{o}}$ and $\boldsymbol{w}_{i,t}^{\mathrm{c}}$ can be bounded such that

$$\begin{aligned} \boldsymbol{w}_{i,t}^{\mathrm{o}} \in \mathcal{W}_i^{\mathrm{o}} :=& \mathcal{W}_i^{\mathrm{c}} \oplus \boldsymbol{L}_i\mathcal{E}_i^{\mathrm{a}} \\ & \oplus \mathrm{co}\left(\left\{\left(\boldsymbol{A}_{\mathrm{d}ii}^{[k]} - \boldsymbol{A}_{\mathrm{d}ii}\left(h^{\mathrm{f}}\right)\right)\mathcal{X}_i \oplus \left(\boldsymbol{B}_{\mathrm{d}ii}^{[k]} - \boldsymbol{B}_{\mathrm{d}ii}\left(h^{\mathrm{f}}\right)\right)\mathcal{U}_i, k \in \mathbb{Z}_{1:n_{\mathrm{pa}}}\right\}\right) \end{aligned} \tag{4.37}$$

$$\boldsymbol{w}_{i,t}^{\mathrm{c}} \in \mathcal{W}_i^{\mathrm{c}} := \bigoplus_{j\in\mathcal{N}_i} \mathrm{co}\left(\left\{\boldsymbol{A}_{\mathrm{d}ij}^{[k]}\mathcal{X}_j \oplus \boldsymbol{B}_{\mathrm{d}ij}^{[k]}\mathcal{U}_j, k \in \mathbb{Z}_{1:n_{\mathrm{pa}}}\right\}\right). \tag{4.38}$$

Note that $\mathcal{W}_i^{\mathrm{c}}$ and $\mathcal{W}_i^{\mathrm{o}}$ are polytopic. If $\boldsymbol{A}_{\mathrm{d}ii}\left(h^{\mathrm{f}}\right) - \boldsymbol{L}_i \boldsymbol{C}_i$ is stable, then there exist RPI sets Ω_i for the error dynamics of the observer. The controller design is straightforward when setting $\boldsymbol{A}_{ii} = \boldsymbol{A}_{\mathrm{d}ii}\left(h^{\mathrm{f}}\right)$ and $\boldsymbol{B}_{ii} = \boldsymbol{A}_{\mathrm{d}ii}\left(h^{\mathrm{f}}\right)$. Note that also Assumptions 4.2.2 and 4.2.3 need to be satisfied.

The stacking up (4.33) and (4.34) for all subsystems $\mathscr{S}_i$, one obtains the augmented system dynamics

$$\boldsymbol{\phi}_{t+1} = \check{\boldsymbol{A}}_{\mathrm{d}}\left(h_t, h^{\mathrm{f}}\right) \boldsymbol{\phi}_t + \check{\boldsymbol{B}}_{\mathrm{d}}\left(h_t, h^{\mathrm{f}}\right) \boldsymbol{u}_t + \check{\boldsymbol{L}} \boldsymbol{e}_t \tag{4.39}$$

with

$$\begin{aligned}
\check{\boldsymbol{A}}_{\mathrm{d}}\left(h_t, h^{\mathrm{f}}\right) &= \begin{pmatrix} \boldsymbol{A}_{\mathrm{d}}(h_t) - \boldsymbol{LC} & \boldsymbol{A}_{\mathrm{d}}(h_t) - \boldsymbol{A}_{\mathrm{dD}}\left(h^{\mathrm{f}}\right) \\ \boldsymbol{LC} & \boldsymbol{A}_{\mathrm{dD}}\left(h^{\mathrm{f}}\right) \end{pmatrix}, \\
\check{\boldsymbol{B}}_{\mathrm{d}}\left(h_t, h^{\mathrm{f}}\right) &= \begin{pmatrix} \boldsymbol{B}_{\mathrm{d}}(h_t) - \boldsymbol{B}_{\mathrm{dD}}\left(h^{\mathrm{f}}\right) \\ \boldsymbol{B}_{\mathrm{dD}}\left(h^{\mathrm{f}}\right) \end{pmatrix}, \\
\boldsymbol{A}_{\mathrm{dD}}\left(h^{\mathrm{f}}\right) &= \mathrm{diag}_{i \in \mathbb{Z}_{1:M}}\left(\boldsymbol{A}_{\mathrm{d}ii}\left(h^{\mathrm{f}}\right)\right), \\
\boldsymbol{B}_{\mathrm{dD}}\left(h^{\mathrm{f}}\right) &= \mathrm{diag}_{i \in \mathbb{Z}_{1:M}}\left(\boldsymbol{B}_{\mathrm{d}ii}\left(h^{\mathrm{f}}\right)\right).
\end{aligned}$$

For ensuring convergence, the following assumption must hold.

Assumption 4.5.1. *For $\check{\boldsymbol{P}} \succ \boldsymbol{0}$, $\check{\boldsymbol{Q}} \succ \boldsymbol{0}$, $\check{\boldsymbol{N}} \succ \boldsymbol{0}$, $0 < \delta < 1$ and $\check{\boldsymbol{S}} = \mathrm{diag}\left(\boldsymbol{S}^{\mathrm{s}}, \boldsymbol{S}^{\mathrm{c}}\right)$, the following matrix inequality is feasible for all $k \in \mathbb{Z}_{1:n_{\mathrm{pa}}}$*

$$\begin{pmatrix} \boldsymbol{\Psi}_{11}^{[k]} & \boldsymbol{\Psi}_{12}^{[k]} \\ * & \boldsymbol{\Psi}_{22}^{[k]} \end{pmatrix} \succeq \boldsymbol{0} \tag{4.40a}$$

$$\boldsymbol{\Psi}_{11}^{[k]} = \check{\boldsymbol{P}} - \check{\boldsymbol{A}}_{\mathrm{d}K}^{[k]}\left(h^{\mathrm{f}}\right)^{\mathrm{T}} \check{\boldsymbol{P}} \check{\boldsymbol{A}}_{\mathrm{d}K}^{[k]}\left(h^{\mathrm{f}}\right) - \sigma^{\mathrm{s}} \check{\boldsymbol{C}}^{\mathrm{T}} \boldsymbol{S}^{\mathrm{s}} \check{\boldsymbol{C}} - \sigma^{\mathrm{c}} \check{\boldsymbol{T}}^{\mathrm{T}} \boldsymbol{S}^{\mathrm{c}} \check{\boldsymbol{T}} - \delta \check{\boldsymbol{Q}} \tag{4.40b}$$

$$\boldsymbol{\Psi}_{12}^{[k]} = \check{\boldsymbol{A}}_{\mathrm{d}K}^{[k]}\left(h^{\mathrm{f}}\right)^{\mathrm{T}} \check{\boldsymbol{P}} \check{\boldsymbol{D}}^{[k]}\left(h^{\mathrm{f}}\right) \tag{4.40c}$$

$$\boldsymbol{\Psi}_{22}^{[k]} = \check{\boldsymbol{S}} - \check{\boldsymbol{D}}^{[k]}\left(h^{\mathrm{f}}\right)^{\mathrm{T}} \check{\boldsymbol{P}} \check{\boldsymbol{D}}^{[k]}\left(h^{\mathrm{f}}\right) - \check{\boldsymbol{N}} \tag{4.40d}$$

where $\check{\boldsymbol{A}}_{\mathrm{d}K}^{[k]}\left(h^{\mathrm{f}}\right) = \check{\boldsymbol{A}}_{\mathrm{d}}^{[k]}\left(h^{\mathrm{f}}\right) + \check{\boldsymbol{B}}_{\mathrm{d}}^{[k]}\left(h^{\mathrm{f}}\right) \check{\boldsymbol{K}}$, $\check{\boldsymbol{D}}^{[k]}\left(h^{\mathrm{f}}\right) = \begin{pmatrix} \check{\boldsymbol{L}} & -\check{\boldsymbol{B}}_{\mathrm{d}}^{[k]}\left(h^{\mathrm{f}}\right) \boldsymbol{K} \end{pmatrix}$ *and*

$$\check{\boldsymbol{A}}_{\mathrm{d}}^{[k]}\left(h^{\mathrm{f}}\right) = \begin{pmatrix} \boldsymbol{A}_{\mathrm{d}}^{[k]} - \boldsymbol{LC} & \boldsymbol{A}_{\mathrm{d}}^{[k]} - \boldsymbol{A}_{\mathrm{dD}}\left(h^{\mathrm{f}}\right) \\ \boldsymbol{LC} & \boldsymbol{A}_{\mathrm{dD}}\left(h^{\mathrm{f}}\right) \end{pmatrix}, \quad \check{\boldsymbol{B}}_{\mathrm{d}}^{[k]}\left(h^{\mathrm{f}}\right) = \begin{pmatrix} \boldsymbol{B}_{\mathrm{d}}^{[k]} - \boldsymbol{B}_{\mathrm{dD}}\left(h^{\mathrm{f}}\right) \\ \boldsymbol{B}_{\mathrm{dD}}\left(h^{\mathrm{f}}\right) \end{pmatrix}.$$

Now the results concerning feasibility and stability can be stated.

Theorem 4.5.2. *Consider the the closed-loop system* (4.26) *and* (4.39). *Assume $\hat{\boldsymbol{x}}_0 \in \mathcal{X}_N$ as well as $\tilde{\boldsymbol{x}}_0 \in \Omega$. Let $\boldsymbol{y}_{i,t}^{\mathrm{ET}}$ and the input sequence $\bar{\boldsymbol{U}}_i^*$ be updated according to* (4.3) *and* (4.24) *$\forall i \in \mathbb{Z}_{1:M}$ and $t \in \mathbb{N}$, respectively. Then $\hat{\boldsymbol{x}}_t \in \mathcal{X}_N$, $\hat{\boldsymbol{x}}_t \in \mathcal{X} \ominus \Omega$, $\boldsymbol{u}_t \in \mathcal{U}$ and $\boldsymbol{x}_t \in \mathcal{X}$ for all $t \in \mathbb{N}$. Moreover, the origin is GES with ROA Φ_N.*

The proof follows the lines of the ones of Theorem 4.4.2 and 4.4.3.

Notably, the design of the state constraints in (4.9a) only ensures constraint satisfaction at the sampling instants t. However, it is desirable that the state constraints are satisfied for all θ, i.e., $\boldsymbol{x}_i(\theta) \in \mathcal{X}_i\ \forall\theta \in \mathbb{R}_{\geq 0}$, $i \in \mathbb{Z}_{1:M}$. One way to achieve this is by imposing tightened constraints on the state of the discrete-time system in a way such that $\boldsymbol{x}_{i,t} \in \mathcal{X}_i^{\text{tight}} \Rightarrow \boldsymbol{x}_i(\theta) \in \mathcal{X}_i$ for all $\theta \in [\theta_t, \theta_{t+1}]$. A sufficient condition is to choose the sets $\mathcal{X}_i^{\text{tight}}$ such that

$$e^{\boldsymbol{A}_{\mathrm{c}}(\theta-\theta_t)} \prod_{i=1}^{M} \mathcal{X}_i^{\text{tight}} \subseteq \mathcal{X} \ominus \int_{\theta_t}^{\theta} e^{\boldsymbol{A}_{\mathrm{c}}s} ds \boldsymbol{B}_{\mathrm{c}} \mathcal{U} \quad \forall\theta \in [\theta_t, \theta_{t+1}] \tag{4.41}$$

holds. Due to the nonlinear dependency on θ, the condition can be hard to check in some cases. One way to verify the condition is to over-approximate the matrices depending on θ by a polytopic set as described above. Note that the observer dynamics (4.5) and thus (4.9c) as well as the input constraints (4.9b) are not affected.

The above approach can also be extended to handle transmission delays $\tau_{i,t}^{\text{S2C}}$ and $\tau_{i,t}^{\text{C2A}}$ in the S2C and C2A channel, respectively, as well as computation delays $\tau_{i,t}^{\text{C}}$ in the controller unit. It is assumed that the sensor unit works time-driven, i.e., it is evaluated at the sampling instants, and the controller as well as actuator unit work event-driven in the sense that they start evaluation the as soon as a new variable is received. The total delay $\tau_{i,t} = \tau_{i,t}^{\text{S2C}} + \tau_{i,t}^{\text{C2A}} + \tau_{i,t}^{\text{C}}$ has to be bounded such that $\tau_{i,t} \in [0, \tau^{\max}]\ \forall i \in \mathbb{Z}_{1:M}$ with $0 \leq \tau^{\max} < h^{\max}$. Following the derivations in [HVG+10] and assuming a constant sampling interval h for reasons of simplicity, an augmented system of the form

$$\chi_{t+1} = \boldsymbol{A}_{\text{da}}(\boldsymbol{\tau}_t)\chi_t + \boldsymbol{B}_{\text{da}}(\boldsymbol{\tau}_t)\boldsymbol{u}_t, \tag{4.42}$$

$$\boldsymbol{y}_t = \boldsymbol{C}_{\text{da}}\chi_t \tag{4.43}$$

can be derived. $\chi_t^{\mathrm{T}} = \begin{pmatrix}\boldsymbol{x}_t^{\mathrm{T}} & \boldsymbol{u}_{t-1}^{\mathrm{T}}\end{pmatrix}^{\mathrm{T}}$ is the the augmented state and

$$\boldsymbol{A}_{\text{da}}(\boldsymbol{\tau}_t) = \begin{pmatrix} e^{\boldsymbol{A}_{\mathrm{c}}h} & \sum_{i=1}^{M}\int_{h-\tau_{i,t}}^{h} e^{\boldsymbol{A}_{\mathrm{c}}s} ds \boldsymbol{B}_{\mathrm{c}i} \\ \boldsymbol{0} & \boldsymbol{0} \end{pmatrix},$$

$$\boldsymbol{B}_{\text{da}}(\boldsymbol{\tau}_t) = \begin{pmatrix} \sum_{i=1}^{M}\int_{0}^{h-\tau_{i,t}} e^{\boldsymbol{A}_{\mathrm{c}}s} ds \boldsymbol{B}_{\mathrm{c}i} \\ \boldsymbol{I} \end{pmatrix},$$

$$\boldsymbol{C}_{\text{da}} = \begin{pmatrix}\boldsymbol{I} & \boldsymbol{0}\end{pmatrix}.$$

$\boldsymbol{B}_{\mathrm{c}i} = \boldsymbol{B}_{\mathrm{c}}\boldsymbol{Y}_i^{\mathrm{T}}\boldsymbol{Y}_i$ where $\boldsymbol{Y}_i \in \mathbb{R}^{m_i \times m}$ is a transformation matrix such that $\boldsymbol{u}_i = \boldsymbol{Y}_i\boldsymbol{u}$. Note that the system and input matrix are depending on the uncertain but bounded parameter $\boldsymbol{\tau}_t = \operatorname{col}_{i\in\mathbb{Z}_{1:M}}(\tau_{i,t})$. The update behavior in the actuator is sketched in Figure 4.3. Note that the time-varying delays which can be different in every subsystem lead to

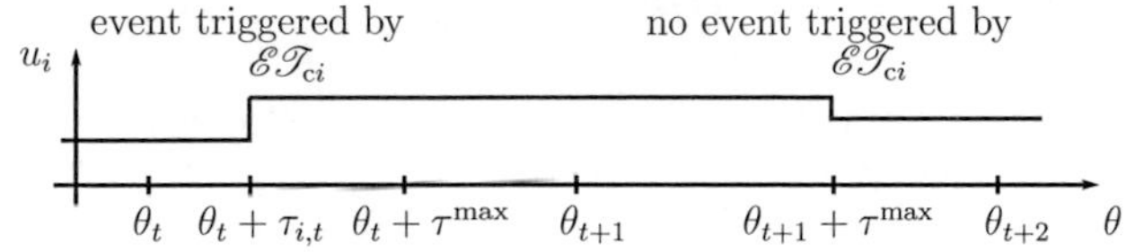

Figure 4.3: The figure illustrates the functional principle for updating the input in actuator $\mathscr{A}_i$. In case of an event, the new input is applied as soon as the new input arrives at the actuator, i.e., at time $\theta_t + \tau_{i,t}$. If there is no event, the actuator waits until $\theta_{t+1} + \tau^{\max}$ to update the plan with a previously received input.

asynchronous updates of the controller as well as actuator units and thus of the inputs of the plants. A controller can be designed exploiting a polytopic over-approximation of the augmented system. The derivation follows the same lines as for the case of time-varying sampling intervals.

The additional uncertainty caused by the time-varying delay and sampling intervals leads to more conservative results. However, if the design is feasible, recursive feasibility, constraint satisfaction and exponential stability of the origin can still be guaranteed in both cases.

Remark 4.5.3. *Note that the event-triggering conditions are checked periodically at every sampling instant. Thus, the inter-event time is at least one sampling interval, i.e., the minimum inter-event time is strictly positive and zeno-behavior cannot occur (see also [HJT12, HDT13]).*

4.6 Illustrative Examples

In this section the control of discrete-time systems in a decentralized set-up is considered first. Then, the derived approach is tested for the control of a sampled-data system with time-varying but bounded sampling intervals for a centralized set-up.

4.6.1 Decentralized Set-up

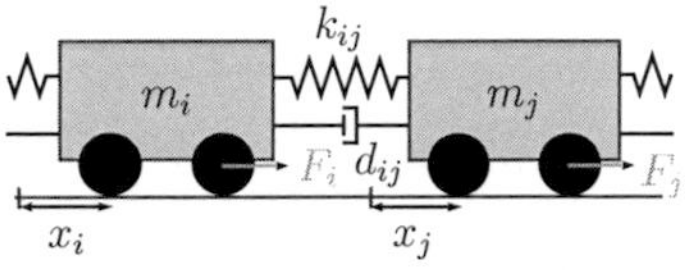

Figure 4.4: Coupled trucks for decentralized set-up

For the decentralized set-up, a modified version of the coupled truck system of [TM17] is investigated with $M = 2$. The set-up is shown in Figure 4.4. The neighbors are defined as $\mathcal{N}_1 = \{2\}$ and $\mathcal{N}_2 = \{1\}$. x_i is the position and $\dot{x}_i$ the velocity of the i-th cart. The input F_i is the force applied to the carts. The continuous-time model of the each subsystem is given by

$$\begin{aligned}\begin{pmatrix}\dot{x}_i\\ \ddot{x}_i\end{pmatrix} &= \begin{pmatrix}0 & 1\\ -\sum_{j\in\mathcal{N}_i}\frac{k_{ij}}{m_i} & -\sum_{j\in\mathcal{N}_i}\frac{d_{ij}}{m_i}\end{pmatrix}\begin{pmatrix}x_i\\ \dot{x}_i\end{pmatrix} + \begin{pmatrix}0\\ 100\end{pmatrix}F_i + \sum_{j\in\mathcal{N}_i}\begin{pmatrix}0 & 1\\ \sum_{j\in\mathcal{N}_i}\frac{k_{ij}}{m_i} & \sum_{j\in\mathcal{N}_i}\frac{d_{ij}}{m_i}\end{pmatrix}\begin{pmatrix}x_j\\ \dot{x}_j\end{pmatrix}\\ y_i &= \begin{pmatrix}1 & 0\\ 0 & 1\end{pmatrix}\begin{pmatrix}x_i\\ \dot{x}_i\end{pmatrix}.\end{aligned} \tag{4.44}$$

The parameters are $m_1 = m_2 = 3\,\mathrm{kg}$, $k_{12} = 0.5\,\mathrm{N/m}$ and $d_{12} = 0.2\,\mathrm{N/m}$. The constraints are $\mathcal{X}_i = \{x_i, \dot{x}_i \in \mathbb{R} \,|\, |x_i| \leq 4\,\mathrm{m}, |\dot{x}_i| \leq 2\,\mathrm{m/s}\}$ and $\mathcal{U}_i = \{F \in \mathbb{R} \,|\, |F| \leq 4\,\mathrm{N}\}$. In the following investigations the discrete-time system of (4.44) is considered. It is obtained by Euler-forward discretization with fixed sampling interval $h = 0.1\,\mathrm{s}$.

The observers are initialized with the first measurement, i.e., $\hat{\boldsymbol{x}}_{i,0} = \boldsymbol{x}_{i,0}$. The observer gain is chosen to $\boldsymbol{L}_i = 0.2 \cdot \boldsymbol{I}$. The controller is designed with a prediction horizon $N = 10$ and $\boldsymbol{K}_i$ is obtained from solving the discrete-time algebraic Riccati equation with $\boldsymbol{Q}_i = \boldsymbol{I}$, $\boldsymbol{R}_i = 0.5$. The weighting matrices are $\boldsymbol{M}_i = \boldsymbol{B}_{ii}^{\mathrm{T}}\boldsymbol{P}_i\boldsymbol{B}_{ii} + \boldsymbol{R}_i$.

First the set-up with event-triggered communication over the S2C channel of the network is considered. The resulting controller is also called (ET-DeMPC) in the sequel. The

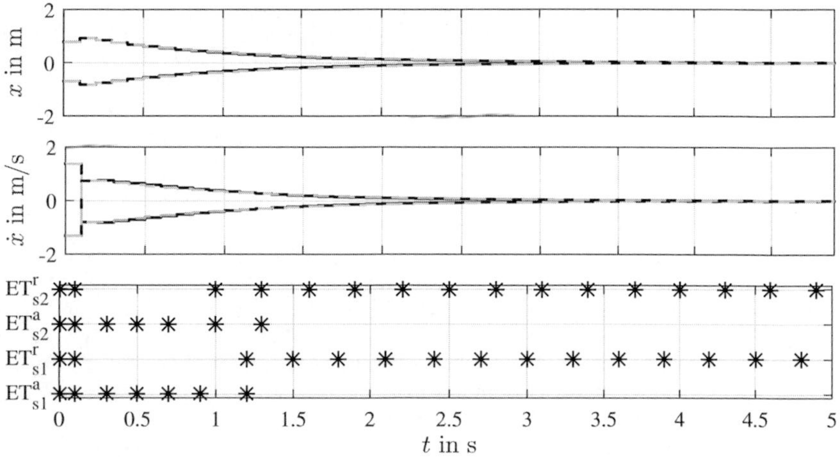

Figure 4.5: Simulation results for the decentralized set-up with trajectory starting from $x_{1,0} = (0.78, 1.37)^{\mathrm{T}}$ and $x_{2,0} = (-0.7 - 1.13)^{\mathrm{T}}$. The evolution of ET-DeMPC and DeMPC are shown in black and gray, respectively. The asterisk indicates if an event is triggered by the corresponding trigger.

absolute trigger sets are $\mathcal{E}_i^{\mathrm{a}} = \{x_i, \dot{x}_i \in \mathbb{R} | |x_i| \leq 0.1\,\mathrm{m}, |\dot{x}_i| \leq 0.1\,\mathrm{m/s}\}$. $\sigma^{\mathrm{s}} = 0.1171$ is obtained by solving an LMI optimization problem maximizing σ^{s} subject to (4.22) for $\delta = 0.1$.

The simulation results are shown in Figure 4.5. One can see that at the beginning the events are triggered by the absolute threshold. When the state approaches the origin, events are triggered by the relative threshold. Moreover, the trajectory of a DeMPC scheme which is designed with the same parameters but $\mathcal{E}_i = \emptyset$ and $\sigma^{\mathrm{s}} = 0$ is shown. The trajectories are almost the same and converge to the origin.

Moreover, controllers with different trigger sets $\mathcal{E}_i^{\mathrm{a}} = \{x_i, \dot{x}_i \in \mathbb{R} | \, \|(x_i, \dot{x}_i)\|_\infty \leq e_i^{\mathrm{a}}\}$ and values of δ are designed. The simulation is run for 100 steps for 100 initial states $\boldsymbol{\phi}_0 \in \{\mathbf{0}\} \times \mathcal{X}_N$ for each controller. The results are shown in Table 4.1. $c_{\mathrm{rel}}^{\mathrm{S2C}}$ and $c_{\mathrm{abs}}^{\mathrm{S2C}}$ are the number of events triggered by the relative and absolute threshold in the sensor unit in relation to the simulation time. $c_{\mathrm{tot}}^{\mathrm{S2C}}$ is the total number of events related to the simulation time. J_{cl} is the closed-loop cost which is defined by $J_{\mathrm{cl}} = \sum_{l=0}^{T_{\mathrm{sim}}} \boldsymbol{x}_t^{\mathrm{T}} \boldsymbol{Q} \boldsymbol{x}_t + \boldsymbol{u}_t^{\mathrm{T}} \boldsymbol{R} \boldsymbol{u}_t$ where $\boldsymbol{Q} = \mathrm{diag}_{i \in \mathbb{Z}_{1:M}}(\boldsymbol{Q}_i)$ and $\boldsymbol{R} = \mathrm{diag}_{i \in \mathbb{Z}_{1:M}}(\boldsymbol{R}_i)$. As expected the number of events triggered by the relative part increases when the values of e_i^{a} decrease. At the same

Table 4.1: Communication effort, cost and ROA for different e_i^{a} and δ in decentralized set-up with event-triggered communication over the S2C channel of the network

	ET-DeMPC						DeMPC
e_1^{a}, e_2^{a}	0.1		0.05		0.01		0
δ	0.1	0.9	0.1	0.9	0.1	0.9	-
$c_{\mathrm{abs}}^{\mathrm{S2C}}$	7.85 %	6.64 %	13.53 %	12.26 %	21.31 %	27.98 %	-
$c_{\mathrm{rel}}^{\mathrm{S2C}}$	27.84 %	44.76 %	26.05 %	42.47 %	29.31 %	34.91 %	-
$c_{\mathrm{tot}}^{\mathrm{S2C}}$	32.60 %	47.04 %	36.41 %	49.67 %	47.13 %	57.85 %	100 %
J_{cl}	33.71	33.70	33.68	33.68	33.66	33.66	33.55
$A_{\mathrm{e}}/A_{\mathrm{DeMPC}}$	60.43 %		68.81 %		75.51 %		100 %

Table 4.2: Communication effort and cost for different e_i^{a} and δ in decentralized set-up with event-triggered communication over the S2C and C2A channels of the network

	ET-DeMPC						DeMPC
e_1^{a}, e_2^{a}	0.1		0.05		0.01		0
δ	0.1	0.9	0.1	0.9	0.1	0.9	-
$c_{\mathrm{tot}}^{\mathrm{S2C}}$	42.15 %	64.11 %	44.92 %	65.75 %	53.81 %	62.21 %	100 %
$c_{\mathrm{tot}}^{\mathrm{C2A}}$	63.55 %	78.17 %	65.48 %	80.09 %	67.89 %	82.09 %	100 %
J_{cl}	33.74	33.73	33.69	33.69	33.67	33.67	33.55

time the cost decreases. While the communication effort caused by the relative trigger increases for larger values of δ, the impact on the performance is small. Furthermore, the performance of the event-triggered controllers compared to the non-event triggered version is also small.

The choice of e_i^{a} also effects the size of the ROA of the controller as can be seen in Figure 4.6. The ratio of the area of the ROAs of the event-triggered controllers to the one of the DeMPC is also listed in Table 4.1.

Next a controller with event-triggered communication over the S2C and C2A channels is designed. The absolute trigger sets are computed by $\mathcal{G}_i^{\mathrm{a},r} = \boldsymbol{A}_{ii}^{-1}\left(\mathcal{F}_i^{r+1} \ominus \mathcal{W}_i\right)$ for $r \in \mathbb{Z}_{1:N-1}$ and $i \in \mathbb{Z}_{1:M}$. Note that with this choice the ROA of the controller is not altered. The results are shown in Table 4.2. Comparing the costs with the costs of the controller with event-triggered communication over the S2C channel only in Table 4.1, one can see that the performance is slightly worse due to the event-triggered update of the inputs. Also the communication effort in the S2C increases. This is mainly caused by the smaller values for σ^{s} that have to be chosen such that LMI (4.27) is feasible.

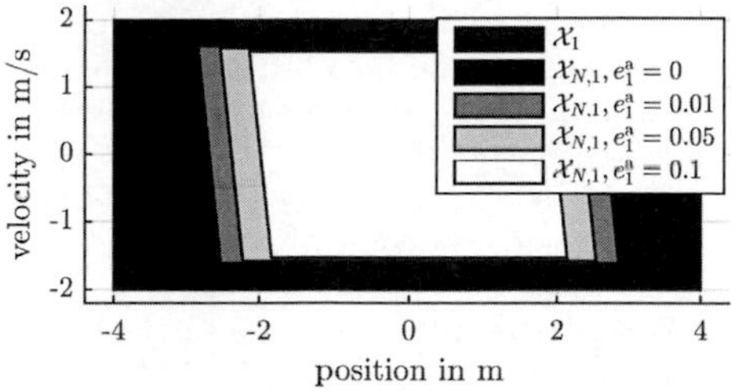

Figure 4.6: Constrained state space $\mathcal{X}_1$ and ROAs $\mathcal{X}_{N,1}$ in $\mathscr{S}_1$ for different values of e_1^a

4.6.2 Centralized Set-up

A ball-and-beam process is considered for demonstration of the method for the centralized set-up ($M = 1$). To improve the readability, the index i is dropped in this subsection. As depicted in Figure 4.7, the beam is mounted at its center on a rotational axis which can be moved by a DC motor connected via a drive belt. The position $x(\theta)$ of the ball is the output of the system and is measured with respect to the center of the beam. The input of the system is the DC motor voltage $u(\theta)$.

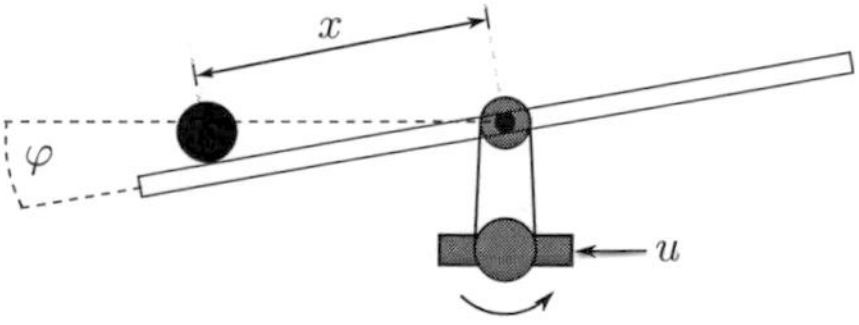

Figure 4.7: Ball-and-beam process for centralized set-up

The continuous-time model for small angles φ is taken from [ÅW97] as

$$\begin{pmatrix} \dot{x} \\ \ddot{x} \end{pmatrix} = \begin{pmatrix} 0 & 1 \\ 0 & 0 \end{pmatrix} \begin{pmatrix} x \\ \dot{x} \end{pmatrix} + \begin{pmatrix} 0 \\ \frac{m \cdot g \cdot k_u}{\Theta / r^2 + m} \end{pmatrix} u, \tag{4.45}$$

$$y = \begin{pmatrix} 1 & 0 \end{pmatrix} \begin{pmatrix} x \\ \dot{x} \end{pmatrix}. \tag{4.46}$$

The control objective is to steer the ball to a reference position $y_{\text{ref}}(\theta)$ on the beam. The parameters are given as $k_u = 0.11\,\text{rad/V}$, $m = 0.11\,\text{kg}$, $\Theta = 10^{-5}\,\text{kg} \cdot \text{m}^2$, $r = 1.5\,\text{cm}$. The process is subject to the constraints $\mathcal{X} = \{x, \dot{x} \in \mathbb{R} |\, |x| \leq 10\,\text{cm}, |\dot{x}| \leq 25\,\text{cm/s}\}$ and

$\mathcal{U} = \{u \in \mathbb{R} |\, |u| \leq 3\,\text{V}\}$. The plant is sampled with a sampling interval $h_t \in [4.9, 5.1]$ms. A discrete-time model is computed using the ZOH method with constant sampling interval $h^{\text{f}} = 5\,\text{ms}$ for the controller. The observer gain is chosen to $\boldsymbol{L} = [0.2, 2]^{\text{T}}$. The controller is designed with a prediction horizon $N = 15$ and $\boldsymbol{K}$ is obtained from solving the discrete-time algebraic Riccati equation with $\boldsymbol{Q} = \text{diag}(1000, 1)$, $\boldsymbol{R} = 0.1$. The weighting matrix $\boldsymbol{M} = \boldsymbol{B}^{\text{T}}\boldsymbol{P}\boldsymbol{B}+\boldsymbol{R}$. The constraints are tightened such that (4.41) holds and thus $\boldsymbol{x}(\theta) \in \mathcal{X}\ \forall\theta \in \mathbb{R}_{\geq 0}$. The absolute trigger value is $\mathcal{E}^{\text{a}} = \{y \in \mathbb{R} |\, |y| \leq 0.5\,\text{mm}\}$. σ^{s} and σ^{c} are computed by solving an LMI problem with (4.40) as constraint for $\delta = 0.9$. The absolute trigger sets are computed by $\mathcal{G}^{\text{a},l} = \boldsymbol{A}^{-1}\left(\mathcal{F}^{l+1} \ominus \mathcal{W}\right)$ for $l \in \mathbb{Z}_{1:N-1}$. For comparison an OB-MPC is designed with the same values as the event-triggered version (called ET-OB-MPC) but with $\mathcal{E} = \emptyset$ and $\mathcal{G}^l = \emptyset$ for $l \in \mathbb{Z}_{0:N}$.

Figure 4.8 shows the simulation results for OB-MPC and ET-OB-MPC for a change of the output reference y_{ref} at $t = 0.2\,\text{s}$ from $0\,\text{cm}$ to $1.2\,\text{cm}$. For both controllers the observer state is initialized after a change of the reference signal with $\hat{\boldsymbol{x}}_0 = \begin{pmatrix} y_{\text{ref},t} & 0 \end{pmatrix}^{\text{T}}$. The controllers steer the outputs to the desired reference despite the uncertainty in the sampling and one can observe a slight difference between the trajectories. For the ET-OB-MPC right after the reference change, many events for S2C communication are triggered due to violation of the absolute trigger. Afterwards events are generated by the relative trigger condition. Looking at the events for updating the input sequence, one can see that the events are triggered just after the reference change with a small delay. After the system is in steady state, no more updates are requested by the absolute but by the relative trigger. Letting the simulation run for 100 randomly chosen feasible reference changes for both set-ups for $T_{\text{sim}} = 100$ steps, the costs are $J_{\text{cl}} = 45.42$ for the OB-MPC and $J_{\text{cl}} = 45.49$ for the ET-OB-MPC scheme.

To quantify the conservatism that is introduced by the time-varying sampling intervals, another comparison is made. A controller with the same data but for fixed sampling interval $h_t = 5\,\text{ms}\ \forall t \in \mathbb{N}$ is designed. Figure 4.9 shows the ROAs of both controllers. One can see that the ROA of the controller considering the varying sampling interval is reduced significantly. Letting the simulation run for 100 randomly chosen feasible reference changes for both set-ups for $T_{\text{sim}} = 100$ steps, the costs are $J_{\text{cl}} = 45.50$ for the time-varying and $J_{\text{cl}} = 45.49$ for the constant sampling interval set-up. One can see that the performance degradation is relatively small.

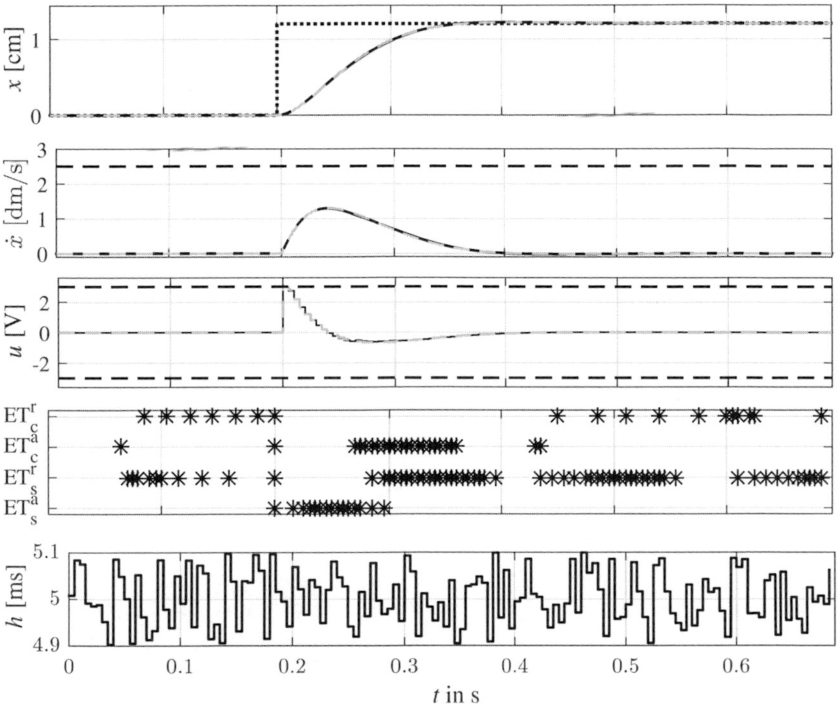

Figure 4.8: Simulation results for the centralized set-up: OB-MPC and ET-OB-MPC are colored gray and black, respectively. The reference signal y^{ref} is dotted and limits are dashed and black. The asterisk indicates if an event is triggered by the corresponding trigger.

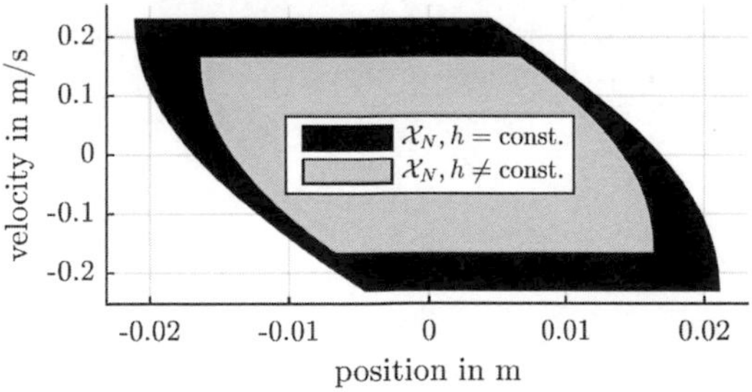

Figure 4.9: Comparison of the ROA of a controller for constant and variable sampling interval

4.7 Summary

In this chapter an OB-MPC scheme for decentralized control of dynamically coupled subsystems with event-triggered communication over a network from the sensors to the controllers is discussed first. The approach employs a combination of event triggers, observers and controllers based on robust MPC. The applied event triggers contain absolute and relative thresholds. While the former are exploited for ensuring recursive feasibility, with the help of the latter attractivity of the origin can be established. The approach is extended to a set-up with additional event-triggered communication over the network from the controllers to the actuators. The applied event triggers use again a combination of absolute and relative thresholds.

Provided that the main assumptions hold, constraint satisfaction, recursive feasibility and exponential stability of the origin can be guaranteed. For the stability analysis of the approach considering both networks, the Lyapunov function derived in Section 3.4 is utilized.

It is outlined how the approach can be transferred to the control of sampled-data systems with bounded time-varying sampling intervals. A polytopic approximation is utilized to over-approximate the nonlinear parametric uncertainty caused by the network-induced imperfection. It is then shown how the controller can be adapted to handle systems with polytopic parametric uncertainty. Furthermore, it is sketched how this approach can be extended to the control of NCSs with bounded time-varying transmission delays in the network. Provided that some more restrictive assumptions (compared to the ones in the discrete-time set-up) hold, constraint satisfaction, recursive feasibility and exponential stability of the origin can still be guaranteed for the case of sampled-data systems with both considered network-induced imperfections.

In the simulation example it is discussed how the selection of the trigger parameters affects the performance, ROA and communication effort for a discrete-time decentralized control set-up. Moreover, the centralized control of a sampled-data system with time-varying transmission intervals is considered in another simulation example. The simulation results of both set-ups show that the communication effort in the network can be reduced significantly compared to the time-triggered equivalents while at the same time the performance degradation is relatively small.

5 Non-iterative DMPC with Event-triggered Communication

In this chapter distributed control of dynamically coupled LTI subsystems subject to decoupled constraints is considered. A non-iterative DMPC scheme is designed where the controllers of the subsystems exchange state information with each other in an event-triggered fashion. The chapter is partly based on [BL18c].

5.1 Problem Set-up

The distributed control set-up depicted in Figure 3.2 is investigated. The dynamics of each subsystem $\mathscr{S}_i$, $i \in \mathbb{Z}_{1:M}$, are given by

$$\mathscr{S}_i : \boldsymbol{x}_{i,t+1} = \boldsymbol{A}_{ii}\boldsymbol{x}_{i,t} + \boldsymbol{B}_{ii}\boldsymbol{u}_{i,t} + \sum_{j \in \mathcal{N}_i} \boldsymbol{A}_{ij}\boldsymbol{x}_{j,t}. \tag{5.1}$$

which is equivalent to the undisturbed state-feedback case of (3.1) with coupling in the dynamics via states. For the sake of convenience, it is assumed that the coupling in the system is bidirectional, i.e., if $j \in \mathcal{N}_i$ then $i \in \mathcal{N}_j$. The states and inputs are subject to the local constraints

$$\boldsymbol{u}_{i,t} \in \mathcal{U}_i \subseteq \mathbb{R}^{m_i}, \quad \boldsymbol{x}_{i,t} \in \mathcal{X}_i \subseteq \mathbb{R}^{n_i}, \tag{5.2}$$

where $\mathcal{X}_i$ and $\mathcal{U}_i$ are compact convex sets containing the origin in their interior. The global dynamics can be described by

$$\mathscr{S} : \boldsymbol{x}_{t+1} = \boldsymbol{A}\boldsymbol{x}_t + \boldsymbol{B}\boldsymbol{u}_t. \tag{5.3}$$

For more information on how (5.1) and (5.3) are related it is referred to Section 3.2. Each subsystem can measure its local state $\boldsymbol{x}_i$ and communicate with the neighboring subsystems over a communication network. Assumptions 3.3.1 and 3.3.2 hold, i.e., the components in the subsystems as well as all the subsystems are synchronized and the communication network is ideal.

The objective is to design a non-iterative DMPC scheme which ensures that the constraints are not violated and the state of the closed-loop converges to the origin while keeping the communication between the subsystems low. To achieve this, each controller is equipped with an event trigger that determines when new state measurements have to be sent to the neighboring subsystems.

The sequel of this chapter is structured as follows. First, an overview of the non-iterative control scheme is given and the required components are introduced in Section 5.2. Then, the theoretical properties of the scheme are analyzed and the main assumptions are stated in Section 5.3. In Section 5.4 the initialization of the algorithm is discussed. A simulation example is given in Section 5.5. The chapter is summarized in Section 5.6.

Note that due to reasons of clarity, the approach does not consider the output-feedback case. However, a combination with OB-MPC, e.g., as described in Section 4.2, can be realized.

5.2 Control Scheme Overview

In this section, first, the controller structure is discussed roughly. Then, each component is introduced. Figure 5.1 depicts the proposed controller structure in each subsystem consisting of a model predictive controller $\mathscr{C}_i$, a predictor $\mathscr{P}_i$ for the local state, predictors $\mathscr{P}_j$ with $j \in \mathcal{N}_i$ for the neighboring states and an event trigger $\mathscr{E}\mathscr{T}_i$. It triggers the sending of updates of $\boldsymbol{x}_i$ to the neighboring subsystems if a threshold is violated. As indicated in the figure, the control scheme requires an initialization. It is discussed in Section 5.4.

For the prediction of the future states in the controller $\mathscr{C}_i$, the following local prediction

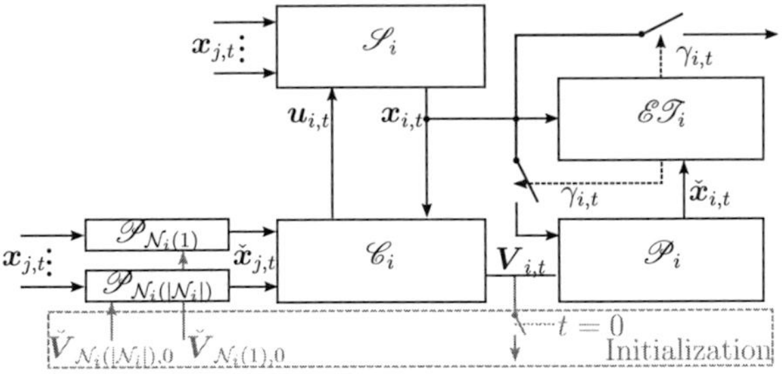

Figure 5.1: Controller set-up in subsystem $\mathscr{S}_i$

model is used

$$\bar{\boldsymbol{x}}_{i,r+1|t} = \boldsymbol{A}_{ii}\bar{\boldsymbol{x}}_{i,r|t} + \boldsymbol{B}_{ii}\bar{\boldsymbol{u}}_{i,r|t} + \sum_{j\in\mathcal{N}_i} \boldsymbol{A}_{ij}\check{\boldsymbol{x}}_{j,r|t}. \tag{5.4}$$

$\bar{\boldsymbol{x}}_{i,r|t}$ and $\bar{\boldsymbol{u}}_{i,r|t}$ are the predicted states and inputs of subsystem $\mathscr{S}_i$ for time $r \in \mathbb{Z}_{t:t+N-1}$ at time t, respectively, and $\check{\boldsymbol{x}}_{j,r|t}$ are the predicted copies of the state of subsystem $\mathscr{S}_j$ known in its neighboring subsystems and thus in $\mathscr{S}_i$. The controller is based on robust MPC methods and uses an input parametrization as described in Section 2.3, i.e., the predicted inputs of subsystem $\mathscr{S}_i$ are given by

$$\bar{\boldsymbol{u}}_{i,r|t} = \boldsymbol{v}_{i,r|t} + \boldsymbol{K}_{ii}\bar{\boldsymbol{x}}_{i,r|t} + \sum_{j\in\mathcal{N}_j} \boldsymbol{K}_{ij}\check{\boldsymbol{x}}_{j,r|t} \tag{5.5}$$

with $\boldsymbol{v}_{i,r|t} \in \mathbb{R}^{m_i}$ for $r \in \mathbb{Z}_{t:t+N-1}$. $\boldsymbol{K}_{ij} \in \mathbb{R}^{m_i\times n_j}$ are local ($j = i$) and coupling ($j \in \mathcal{N}_i$) feedback matrices. Because $\boldsymbol{K}_{ij} = \boldsymbol{0}$ for all $j \notin \mathcal{N}_i \cup \{i\}$, the structure of the global feedback matrix $\boldsymbol{K} = (\boldsymbol{K}_{ij})$ is identical to the structure in $\boldsymbol{A}$. This means that for the computation of control law (5.5), no additional communication path is required. The local prediction model (5.4) can now be rewritten as

$$\bar{\boldsymbol{x}}_{i,r+1|t} = \boldsymbol{F}_{ii}\bar{\boldsymbol{x}}_{i,r|t} + \boldsymbol{B}_{ii}\boldsymbol{v}_{i,r|t} + \sum_{j\in\mathcal{N}_i} \boldsymbol{F}_{ij}\check{\boldsymbol{x}}_{j,r|t}. \tag{5.6}$$

with $\boldsymbol{F}_{ij} = \boldsymbol{A}_{ij} + \boldsymbol{B}_{ii}\boldsymbol{K}_{ij}$ $j \in \mathcal{N}_i \cup \{i\}$. Moreover, the global matrix $\boldsymbol{F} = \boldsymbol{A} + \boldsymbol{BK}$ is defined. Note that the matrix $\boldsymbol{F}$ has the same structure as $\boldsymbol{A}$ and has to fulfill the following main assumption.

Assumption 5.2.1. *The matrix $\boldsymbol{F}$ is stable.*

For computation of the future states, an initial state $\bar{\boldsymbol{x}}_{i,t|t} = \boldsymbol{x}_{i,t}$, i.e., the local measurement, and the sequence $\boldsymbol{v}_{i,r|t}$, which is computed by the DMPC controller, are required. The states of the neighboring subsystems $\check{\boldsymbol{x}}_{j,r|t}$ are generated using the approximated neighbor model

$$\check{\boldsymbol{x}}_{j,r+1|t} = \boldsymbol{F}_{jj}\check{\boldsymbol{x}}_{j,r|t} + \boldsymbol{B}_{jj}\check{\boldsymbol{v}}_{j,r|t},\ \forall r \in \mathbb{Z}_{t:t+N-1},\ \forall j \in \mathcal{N}_i \tag{5.7}$$

where $\check{\boldsymbol{v}}_j \in \mathbb{R}^{m_j}$ is a copy of $\boldsymbol{v}_j$. Having an initial value $\check{\boldsymbol{x}}_{j,t|t} = \check{\boldsymbol{x}}_{j,t}$ and a sequence $\check{\boldsymbol{V}}_{j,t} = \left[\check{\boldsymbol{v}}^{\mathrm{T}}_{j,t|t}, \dots, \check{\boldsymbol{v}}^{\mathrm{T}}_{j,t+N-1|t}\right]^{\mathrm{T}}$, the states $\check{\boldsymbol{x}}_{j,r}$ can be computed in the neighboring subsystems of subsystem $\mathscr{S}_j$. The initial state values $\check{\boldsymbol{x}}_{j,t}$ are generated using the predictors $\mathscr{P}_j$ for the neighboring states. The input sequence $\check{\boldsymbol{V}}_{j,t}$ is computed at time $t = 0$, sent to the neighboring systems and updated in every time step $t > 0$. More details are given in the sequel of the chapter. Note that (5.7) does not consider the coupling to the neighbors of subsystem $\mathscr{S}_j$ in order to limit the dimension of the local models. However, this means that there is an error in the prediction model.

5.2.1 Event Trigger and Predictor

The update rule in the event trigger $\mathscr{ET}_i$ is

$$\mathscr{ET}_i : \gamma_{i,t} = \begin{cases} 0 & \text{if } \boldsymbol{x}_{i,t} - (\boldsymbol{A}_{ii}\check{\boldsymbol{x}}_{i,t-1} + \boldsymbol{B}_{ii}\check{\boldsymbol{u}}_{i,t-1}) \in \mathcal{E}_i(\boldsymbol{x}_{i,t}) \\ 1 & \text{if } \boldsymbol{x}_{i,t} - (\boldsymbol{A}_{ii}\check{\boldsymbol{x}}_{i,t-1} + \boldsymbol{B}_{ii}\check{\boldsymbol{u}}_{i,t-1}) \notin \mathcal{E}_i(\boldsymbol{x}_{i,t}) \end{cases} \tag{5.8}$$

where $\gamma_i \in \mathbb{Z}_{0:1}$ is a binary variable indicating if an event is triggered or not and $\check{\boldsymbol{u}}_{i,t} = \boldsymbol{K}_{ii}\check{\boldsymbol{x}}_{i,t} + \check{\boldsymbol{v}}_{i,t}$. The state of the predictor $\mathscr{P}_i$ is called $\check{\boldsymbol{x}}_{i,t}$. The predictor is running in the local subsystem $\mathscr{S}_i$ and all of its neighboring subsystems $\mathscr{S}_j$ with $j \in \mathcal{N}_i$ meaning that the state $\check{\boldsymbol{x}}_{i,t}$ is available in all subsystems $\mathscr{S}_j$ with $j \in \mathcal{N}_i \cup \{i\}$. The state $\check{\boldsymbol{x}}_{i,t}$ is updated according to

$$\mathscr{P}_i : \check{\boldsymbol{x}}_{i,t} = \begin{cases} \boldsymbol{A}_{ii}\check{\boldsymbol{x}}_{i,t-1} + \boldsymbol{B}_{ii}\check{\boldsymbol{u}}_{i,t-1} & \text{if } \gamma_{i,t} = 0 \\ \boldsymbol{x}_{i,t} & \text{if } \gamma_{i,t} = 1 \end{cases} \quad \forall i \in \mathbb{Z}_{1:M}. \tag{5.9}$$

In the case of no update, the same model in (5.9) is used as in the prediction model (5.7). In the case of an update, the state measurement is applied. This means that only in the case of an event $\boldsymbol{x}_{i,t}$ needs to be sent to the neighboring subsystems, otherwise $\check{\boldsymbol{x}}_{i,t}$ is generated by the predictors locally. Note that the predictors for constructing the same state $\check{\boldsymbol{x}}_{i,t}$ generate the same values independent from the subsystem they are located in, i.e., in $\mathscr{S}_j$ with $j \in \mathcal{N}_i \cup \{i\}$, as they run identical models with the same values for $\check{v}_i$ and are updated by the event-triggers in the same way.

The error $\boldsymbol{e}_i$ caused by the event-triggered communication, i.e., the difference between the state of the predictor and the actual state, is bounded by

$$\boldsymbol{e}_{i,t} = \boldsymbol{x}_{i,t} - \check{\boldsymbol{x}}_{i,t} \in \mathcal{E}_i(\boldsymbol{x}_{i,t}) \ \forall i \in \mathbb{Z}_{1:M},\ t \in \mathbb{N}. \tag{5.10}$$

The triggering set $\mathcal{E}_i(\boldsymbol{x}_i) \subseteq \mathbb{R}^{n_i}$ is given by

$$\mathcal{E}_i(\boldsymbol{x}_i) = \mathcal{E}_i^{\mathrm{a}} \cap \mathcal{E}_i^{\mathrm{r}}(\boldsymbol{x}_i). \tag{5.11}$$

It is the intersection of the relative triggering set

$$\mathcal{E}_i^{\mathrm{r}}(\boldsymbol{x}_i) = \left\{\boldsymbol{z}_i \in \mathbb{R}^{n_i} \middle| \boldsymbol{z}_i^{\mathrm{T}}\boldsymbol{S}_i\boldsymbol{z}_i \leq \sigma\boldsymbol{x}_i^{\mathrm{T}}\boldsymbol{S}_i\boldsymbol{x}_i\right\} \subseteq \mathbb{R}^{n_i} \tag{5.12}$$

and the absolute triggering set $\mathcal{E}_i^{\mathrm{a}} \subseteq \mathbb{R}^{n_i}$ which is convex and compact as well as $\boldsymbol{0} \in \mathcal{E}_i^{\mathrm{a}}$ holds. $\boldsymbol{S}_i \in \mathbb{R}^{n_i \times n_i}, \boldsymbol{S}_i \succ \boldsymbol{0}$ is a scaling matrix and $\sigma \in \mathbb{R}_{\geq 0}$ a tuning parameter to adjust the communication effort. More information on their design is given in Section 5.3. As shown in the sequel of the chapter, with the help of the absolute triggering set recursive feasibility can be guaranteed. Convergence of the closed-loop to the origin can be ensured by exploiting the relative part.

5.2.2 Error Propagation

Since the prediction models (5.6) and (5.7) use approximations for the neighboring subsystems and event-triggered communication is applied, there is a deviation to the actual evolution of the plant. This difference can be interpreted as an artificial disturbance acting on the prediction models and thus a robust control design is applied to ensure recursive feasibility. For the design, a bound on the disturbance $\boldsymbol{w}_{i,t} = \boldsymbol{x}_{i,t+1} - \bar{\boldsymbol{x}}_{i,t+1|t}$, i.e., the difference of the predicted state at time t for time $t+1$ to the actual state at time $t+1$, is required. Subtracting (5.4) from (5.1) considering $\boldsymbol{u}_{i,t} = \bar{\boldsymbol{u}}_{i,t}$ and recalling that $\boldsymbol{x}_{i,t} = \bar{\boldsymbol{x}}_{i,t}$ as well as (5.10), the disturbance can be written as and bounded by

$$\boldsymbol{w}_{i,t} = \boldsymbol{x}_{i,t+1} - \bar{\boldsymbol{x}}_{i,t+1|t} = \sum_{j \in \mathcal{N}_i} \boldsymbol{A}_{ij}(\boldsymbol{x}_{j,t} - \check{\boldsymbol{x}}_{j,t}) \in \mathcal{W}_i \tag{5.13}$$

where $\mathcal{W}_i = \bigoplus_{j \in \mathcal{N}_i} \boldsymbol{A}_{ij} \mathcal{E}_j^{\mathrm{a}}$.

Now the artificial disturbance $\check{\boldsymbol{w}}_{j,t}$ acting on the neighbor model (5.7) is discussed. In case of an event at time $t+1$, the neighbor model is updated with $\boldsymbol{x}_{j,t+1}$ and thus by subtracting (5.7) from (5.1) one obtains

$$\check{\boldsymbol{w}}_{j,t} = \boldsymbol{x}_{j,t+1} - \check{\boldsymbol{x}}_{j,t+1|t} \tag{5.14}$$

$$= \boldsymbol{F}_{jj}\left(\boldsymbol{x}_{j,t} - \check{\boldsymbol{x}}_{j,t}\right) + \boldsymbol{B}_{jj}\Delta \boldsymbol{v}_{j,t} + \sum_{k \in \mathcal{N}_j} \left(\boldsymbol{F}_{jk}\boldsymbol{x}_{k,t} - \boldsymbol{B}_{jj}\boldsymbol{K}_{jk}\left(\boldsymbol{x}_{k,t} - \check{\boldsymbol{x}}_{k,t}\right)\right). \tag{5.15}$$

The difference $\boldsymbol{x}_{k,t} - \check{\boldsymbol{x}}_{k,t} \in \mathcal{E}_k^{\mathrm{a}}$ $\forall k \in \mathcal{N}_j \cup \{j\}$ see (5.10). $\Delta \boldsymbol{v}_{j,t} = \boldsymbol{v}_{j,t} - \check{\boldsymbol{v}}_{j,t}$ is the difference of the communicated value $\check{\boldsymbol{v}}_{j,t}$ and the actual value $\boldsymbol{v}_{j,t}$. In order to bound $\check{\boldsymbol{w}}_{j,t}$, $\Delta \boldsymbol{v}_{i,t}$ needs to be bounded $\forall i \in \mathbb{Z}_{1:M}$, i.e.,

$$\Delta \boldsymbol{v}_{i,t} \in \mathcal{V}_i \tag{5.16}$$

where $\mathcal{V}_i \subseteq \mathbb{R}^{m_i}$ are compact convex sets with $\boldsymbol{0} \in \mathcal{V}_i$. As shown in the next section, $\Delta \boldsymbol{v}_{i,t} \in \mathcal{V}_i$ is included as a constraint in the optimization problem meaning that the local controller is only allowed to deviate $\boldsymbol{v}_{i,r}$ from the initially communicated value $\check{\boldsymbol{v}}_{i,r}$ within some region $\mathcal{V}_i$. So for the case of an event at time $t+1$, $\check{\boldsymbol{w}}_{j,t} \in \check{\mathcal{W}}_j$ where

$$\check{\mathcal{W}}_j = \boldsymbol{F}_{jj}\mathcal{E}_j^{\mathrm{a}} \oplus \boldsymbol{B}_{jj}\mathcal{V}_j \oplus \bigoplus_{k \in \mathcal{N}_j} \left(\boldsymbol{F}_{jk}\mathcal{X}_k \oplus \left(-\boldsymbol{B}_{jj}\boldsymbol{K}_{jk}\right)\mathcal{E}_k^{\mathrm{a}}\right). \tag{5.17}$$

Note that the sets $\check{\mathcal{W}}_j$ are compact. In case of no event at $t+1$, the disturbance $\check{\boldsymbol{w}}_{j,t} = \boldsymbol{0}$ $\forall j \in \mathcal{N}_i$ as the predictor (5.9) as well as the neighbor model (5.7) generate the same state. Since $\boldsymbol{0} \in \check{\mathcal{W}}_j$, the case of no event is also covered in (5.17).

Considering (5.17), one can observe that the disturbance is caused for three reasons. The first and forth disturbance term are caused by the event-triggered communication style. When the size of the event-trigger set $\mathcal{E}_i^{\mathrm{a}}$ is reduced, the disturbance term becomes smaller but the number of events is expected to increase. The fact that the input $\boldsymbol{v}_i$ can be different compared to the previously communicated value is considered in the second part. Note that this constraint is also referred to as a consistency constraint in the literature. As outlined in the introduction, this is a common way in non-iterative schemes to lower the uncertainty imposed on the other subsystems. Choosing larger values for $\mathcal{V}_i$ gives more freedom to the local controllers but increases the size of $\check{\mathcal{W}}_j$. The third part is due to neglecting the influence of the neighbors' neighbors in the approximated neighbor model (5.7). It can be altered by designing the controller gain components $\boldsymbol{K}_{jk}$ such that the influence of $\boldsymbol{F}_{jk}\mathcal{X}_k$ on $\check{\mathcal{W}}_j$ is minimal. In some cases it is even possible to select $\boldsymbol{F}_{jk} = \mathbf{0}$. This procedure is known in literature as perfect matching or perfect decoupling [Šil91]. Because in practice the coupling terms might not be known exactly or $\boldsymbol{B}_{jk}$ might not have full rank, in general it is not possible to select $\boldsymbol{F}_{jk} = \mathbf{0}$.

5.2.3 Model Predictive Controller

The controller in every subsystem utilizes the models (5.6) and (5.7). Due to the imperfect neighbor models and the event-triggered communication, a robust controller design is required to ensure recursive feasibility. In this chapter, the robust MPC approach discussed in Section 2.3 is applied. The local optimization problem to be solved in every controller $\mathscr{C}_i$ is of the form

$$\underset{\boldsymbol{V}_{i,t}}{\text{minimize}} \quad J_{N,i}\left(\boldsymbol{V}_{i,t}\right) \tag{5.18a}$$

$$\text{subject to} \quad \bar{\boldsymbol{x}}_{i,t|t} = \boldsymbol{x}_{i,t} \text{ and } \check{\boldsymbol{x}}_{j,t|t} = \check{\boldsymbol{x}}_{j,t}\, \forall j \in \mathcal{N}_i \tag{5.18b}$$

$$\bar{\boldsymbol{x}}_{i,t+N|t} \in \bar{\mathcal{X}}_i^{\mathrm{f}} \tag{5.18c}$$

$$\text{and } \forall r \in \mathbb{Z}_{0:N-1}:$$

$$\bar{\boldsymbol{x}}_{i,r+1|t} = \boldsymbol{F}_{ii}\bar{\boldsymbol{x}}_{i,r|t} + \boldsymbol{B}_{ii}\boldsymbol{v}_{i,r|t} + \sum_{j\in\mathcal{N}_i} \boldsymbol{F}_{ij}\check{\boldsymbol{x}}_{j,r|t} \tag{5.18d}$$

$$\check{\boldsymbol{x}}_{j,r+1|t} = \boldsymbol{F}_{jj}\check{\boldsymbol{x}}_{j,r|t} + \boldsymbol{B}_{jj}\check{\boldsymbol{v}}_{j,r|t} \tag{5.18e}$$

$$\bar{\boldsymbol{x}}_{i,r|t} \in \bar{\mathcal{X}}_i^r, \tag{5.18f}$$

$$\boldsymbol{v}_{i,r|t} + \boldsymbol{K}_{ii}\bar{\boldsymbol{x}}_{i,r|t} + \sum_{j\in\mathcal{N}_i} \boldsymbol{K}_{ij}\check{\boldsymbol{x}}_{j,r|t} \in \bar{\mathcal{U}}_i^r \tag{5.18g}$$

$$\boldsymbol{v}_{i,r|t} - \check{\boldsymbol{v}}_{i,r|t} \in \mathcal{V}_i. \tag{5.18h}$$

In the optimization problem the prediction models (5.6) and (5.7) are used in (5.18d+e). They are updated with the local measurements and the predictor state for the neighboring subsystems in (5.18b). $\boldsymbol{V}_{i,t} = \left[\boldsymbol{v}_{i,t}^{\mathrm{T}}, \ldots, \boldsymbol{v}_{i,t+N-1}^{\mathrm{T}}\right]^{\mathrm{T}}$ is the local input sequence.

The constraints (5.18h) assure that the deviation of $\boldsymbol{v}_{i,r}$ to $\check{\boldsymbol{v}}_{i,r}$ is bounded as required by (5.16). The local cost function (5.18a) is given by

$$J_{N,i}\left(\boldsymbol{V}_{i,t}\right) = \sum_{r=t}^{t+N-1} \boldsymbol{v}_{i,r}^{\mathrm{T}} \boldsymbol{M}_i \boldsymbol{v}_{i,r} \tag{5.19}$$

where $\boldsymbol{M}_i \in \mathbb{R}^{m_i \times m_i}$, $\boldsymbol{M}_i \succ \boldsymbol{0}$ is the weighting matrix. The constraints (5.2) are included in (5.18f+g) and are tightened such that

$$\begin{aligned}
\bar{\mathcal{X}}_i^r &= \bar{\mathcal{X}}_i^{r-1} \ominus \bar{\mathcal{W}}_i^r, & \bar{\mathcal{X}}_i^0 &= \mathcal{X}_i & (5.20a)\\
\bar{\mathcal{U}}_i^r &= \bar{\mathcal{U}}_i^{r-1} \ominus \boldsymbol{K}_{ii}\bar{\mathcal{W}}_i^r \ominus \bigoplus_{j\in\mathcal{N}_i} \boldsymbol{K}_{ij}\check{\mathcal{W}}_j^r, & \bar{\mathcal{U}}_i^0 &= \mathcal{U}_i & (5.20b)\\
\bar{\mathcal{W}}_i^r &= \boldsymbol{F}_{ii}\bar{\mathcal{W}}_i^{r-1} \oplus \bigoplus_{j\in\mathcal{N}_i} \boldsymbol{F}_{ij}\check{\mathcal{W}}_j^{r-1}, & \bar{\mathcal{W}}_i^0 &= \mathcal{W}_i & (5.20c)\\
\check{\mathcal{W}}_j^r &= \boldsymbol{F}_{jj}\check{\mathcal{W}}_j^{r-1}, & \check{\mathcal{W}}_j^0 &= \check{\mathcal{W}}_j. & (5.20d)
\end{aligned}$$

for $r \in \mathbb{Z}_{1:N}$. $\bar{\mathcal{X}}_i^{\mathrm{f}} \subseteq \mathbb{R}^{n_i}$ is the terminal set for which the following main assumption has to hold.

Assumption 5.2.2. *There exist sets $\check{\mathcal{X}}_i^{\mathrm{f}} \subseteq \mathbb{R}^{n_i}$, $\bar{\check{\mathcal{X}}}_i^{\mathrm{f}} \subseteq \mathbb{R}^{n_i}$, $\mathcal{X}_i^{\mathrm{f}} \subseteq \mathbb{R}^{n_i}$ and $\bar{\mathcal{X}}_i^{\mathrm{f}}$ $\forall i \in \mathbb{Z}_{1:M}$ such that*

$$\begin{aligned}
&\check{\boldsymbol{x}}_i \in \check{\mathcal{X}}_i^{\mathrm{f}}, \check{\boldsymbol{w}}_i \in \check{\mathcal{W}}_i^N : \boldsymbol{F}_{ii}\check{\boldsymbol{x}}_i + \check{\boldsymbol{w}}_i \in \check{\mathcal{X}}_i^{\mathrm{f}}, & (5.21a)\\
&\bar{\check{\mathcal{X}}}_i^{\mathrm{f}} = \check{\mathcal{X}}_i^{\mathrm{f}} \ominus \check{\mathcal{W}}_i^N & (5.21b)\\
&\boldsymbol{x}_i \in \mathcal{X}_i^{\mathrm{f}}, \boldsymbol{w}_i \in \bar{\mathcal{W}}_i^N, \check{\boldsymbol{x}}_j \in \check{\mathcal{X}}_j^{\mathrm{f}}\ \forall j \in \mathcal{N}_i : & (5.21c)\\
&\boldsymbol{F}_{ii}\boldsymbol{x}_i + \sum_{j\in\mathcal{N}_i} \boldsymbol{F}_{ij}\check{\boldsymbol{x}}_j + \boldsymbol{w}_i \in \mathcal{X}_i^{\mathrm{f}}, & (5.21d)\\
&\mathcal{X}_i^{\mathrm{f}} \subseteq \bar{\mathcal{X}}_i^N, & (5.21e)\\
&\boldsymbol{K}_{ii}\mathcal{X}_i^{\mathrm{f}} \oplus \bigoplus_{j\in\mathcal{N}_i} \boldsymbol{K}_{ij}\check{\mathcal{X}}_j^{\mathrm{f}} \subseteq \bar{\mathcal{U}}_i^N, & (5.21f)\\
&\bar{\mathcal{X}}_i^{\mathrm{f}} = \mathcal{X}_i^{\mathrm{f}} \ominus \bar{\mathcal{W}}_i^N. & (5.21g)
\end{aligned}$$

Assumption 5.2.2 requires that $\bar{\mathcal{X}}_i^{\mathrm{f}}$ and $\bar{\check{\mathcal{X}}}_j^{\mathrm{f}}$ for $j \in \mathcal{N}_i$ form a structured RPI sets for the dynamics in (5.21a+d) for $i \in \mathbb{Z}_{1:M}$. Note that the sets $\bar{\check{\mathcal{X}}}_i^{\mathrm{f}}$ are not required in (5.18). However, it has to be considered that $\check{\boldsymbol{x}}_{i,N|0} \in \bar{\check{\mathcal{X}}}_i^{\mathrm{f}}$ holds when the initial sequences $\check{\boldsymbol{V}}_{i,0}$ are computed (see Section 5.4).

The local control input is

$$\boldsymbol{u}_{i,t} = \boldsymbol{v}_{i,t|t}^{*} + \boldsymbol{K}_{ii}\boldsymbol{x}_{i,t} + \sum_{j\in\mathcal{N}_j} \boldsymbol{K}_{ij}\check{\boldsymbol{x}}_{j,t},\ \forall i \in \mathbb{Z}_{1:M} \tag{5.22}$$

Algorithm 1 Non-iterative DMPC Algorithm for each subsystem $\mathscr{S}_i$:

Require: $\check{\boldsymbol{x}}_{j,t-1}$ and $\check{\boldsymbol{V}}_{j,t}$ $j \in \mathcal{N}_i \cup \{i\}$
1: Measure $\boldsymbol{x}_{i,t}$
2: Evaluate $\mathscr{E}\mathscr{T}_i$ according to (5.8). If $\gamma_{i,t} = 1$, send $\boldsymbol{x}_{i,t}$ to neighbors $j \in \mathcal{N}_i$
3: Update predictors $\mathscr{P}_j$ according to (5.9) for $j \in \mathcal{N}_i \cup \{i\}$
4: Solve optimization problem (5.18)
5: Update the input sequences $\check{\boldsymbol{V}}_{j,t+1} = \left[\check{\boldsymbol{v}}_{j,t+1|t}^{\mathrm{T}}, \dots, \check{\boldsymbol{v}}_{j,t+N-1|t}^{\mathrm{T}}, \mathbf{0}^{\mathrm{T}}\right]^{\mathrm{T}}$ for $j \in \mathcal{N}_i \cup \{i\}$
6: Compute $\boldsymbol{u}_{i,t}$ according to (5.22) and apply $\boldsymbol{u}_{i,t}$ to plant

and $\boldsymbol{v}_{i,t|t}^*$ is the first entry of the sequence $\boldsymbol{V}_{i,t}^*$ obtained as the solution of (5.18). The input applied to the global dynamics (5.3) is

$$\boldsymbol{u}_t = \mathrm{col}_{i \in \mathbb{Z}_{1:M}} \left(\boldsymbol{u}_{i,t}\right). \tag{5.23}$$

Algorithm 1 summarizes the steps of the proposed method. After measuring the current state in step 1, in step 2 and 3 the event trigger and predictors are evaluated. The DMPC problem is solved in step 4. In step 5 the sequences $\check{\boldsymbol{V}}_{j,t}$ of the neighboring subsystems and the exchanged sequence $\check{\boldsymbol{V}}_{i,t}$ are updated by shifting the old sequence and extending it with the zero vector.

Remark 5.2.3. *As pointed out in Section 2.3, the assumption of RPI terminal sets is standard in many robust MPC approaches. The additional structural assumption on the set can also be found in several non-iterative DMPC approaches, for instance, in [FS12] or [HT16].*

5.3 Analysis

Before the main theoretical results are stated, the following definitions are required: The vectors $\boldsymbol{V}_t$ and $\check{\boldsymbol{V}}_t$ comprises all sequences $\boldsymbol{V}_{i,t}$ and $\check{\boldsymbol{V}}_{i,t}$ for $i \in \mathbb{Z}_{1:M}$, respectively. Furthermore,

$$\begin{aligned} \mathcal{X}_N = \{ & \boldsymbol{x} \in \mathbb{R}^n | \text{ for } \boldsymbol{x}_i = \boldsymbol{W}_i \boldsymbol{x}\ \check{\boldsymbol{x}}_j = \boldsymbol{W}_j \boldsymbol{x}\ \forall j \in \mathcal{N}_i \\ & \exists\, \boldsymbol{V}_i,\ \check{\boldsymbol{V}}_i \text{ such that (5.18) is feasible } \forall i \in \mathbb{Z}_{1:M} \}, \end{aligned} \tag{5.24}$$

is the set of feasible states,

$$\begin{aligned} \check{\mathcal{D}}_N(\boldsymbol{x}) = \{ & \check{\boldsymbol{V}} \in \mathbb{R}^{m \cdot N} | \boldsymbol{x} \in \mathcal{X}_N \text{ and for } \boldsymbol{x}_i = \boldsymbol{W}_i \boldsymbol{x},\ \check{\boldsymbol{x}}_j = \boldsymbol{W}_j \boldsymbol{x} \\ & \forall j \in \mathcal{N}_i\ \exists\, \boldsymbol{V}_i, \text{ s.t. (5.18) is feasible } \forall i \in \mathbb{Z}_{1:M} \} \end{aligned} \tag{5.25}$$

is the set of feasible sequences $\check{\boldsymbol{V}}$ and

$$\mathcal{D}_N\left(\boldsymbol{x},\check{\boldsymbol{V}}\right)=\left\{\boldsymbol{V}\in\mathbb{R}^{m\cdot N}|\check{\boldsymbol{V}}\in\check{\mathcal{D}}_N\left(\boldsymbol{x}\right)\right\} \tag{5.26}$$

is the set of feasible sequences $\boldsymbol{V}$.

Next, the results concerning recursive feasibility can be stated.

Theorem 5.3.1. *Consider system* (5.3) *and assume that* $\boldsymbol{x}_0\in\mathcal{X}_N$*, an initial sequence* $\check{\boldsymbol{V}}_0\in\check{\mathcal{D}}_N\left(\boldsymbol{x}_0\right)$ *is given, the control law* (5.22) *and* $\check{\boldsymbol{V}}_t$ *are determined by Algorithm 1 as well as the event-triggers and predictors are evaluated according to Algorithm 1, then* $\boldsymbol{x}_t\in\mathcal{X}_N$ *and* $\check{\boldsymbol{V}}_t\in\check{\mathcal{D}}_N\left(\boldsymbol{x}_t\right)$ *for all* $t\in\mathbb{N}$.

Proof: The proof can be found in Appendix B.3. □

Next, convergence of the closed-loop system is discussed (in the sense of attractivity of the origin). The procedure is based on the method presented in Section 3.4. For this purpose, the function

$$V_N^*\left(\boldsymbol{x}_t,\check{\boldsymbol{V}}_t\right)=V^{\mathrm{f}}\left(\boldsymbol{x}_t\right)+\nu J_N^*\left(\boldsymbol{x}_t,\check{\boldsymbol{V}}_t\right) \tag{5.27}$$

where $V^{\mathrm{f}}\left(\boldsymbol{x}\right)=\boldsymbol{x}^{\mathrm{T}}\boldsymbol{P}\boldsymbol{x}$, $\boldsymbol{P}\in\mathbb{R}^{n\times n}$, $\nu\in\mathbb{R}_{>0}$ and $J_N^*\left(\boldsymbol{x}_t,\check{\boldsymbol{V}}_t\right)=\min_{\boldsymbol{V}_t\in\mathcal{D}_N(\boldsymbol{x}_t,\check{\boldsymbol{V}}_t)}J_N\left(\boldsymbol{V}_t\right)$, i.e., the optimal value of the global cost function

$$J_N\left(\boldsymbol{V}_t\right)=\sum_{i=1}^{M}J_{N,i}\left(\boldsymbol{V}_{i,t}\right)=\sum_{r=t}^{t+N-1}\boldsymbol{v}_r^{\mathrm{T}}\boldsymbol{M}\boldsymbol{v}_r \tag{5.28}$$

with $\boldsymbol{M}=\operatorname{diag}_{i\in\mathbb{Z}_{1:M}}\left(\boldsymbol{M}_i\right)$ is defined. System (5.3) under control law (5.23) can be recast as the closed-loop system

$$\boldsymbol{x}_{t+1}=\boldsymbol{F}\boldsymbol{x}_t+\boldsymbol{B}\boldsymbol{v}_t+\boldsymbol{B}\boldsymbol{K}_{\mathrm{C}}\boldsymbol{e}_t \tag{5.29}$$

where $\boldsymbol{K}_{\mathrm{C}}=\boldsymbol{K}-\operatorname{diag}_{i\in\mathbb{Z}_{1:M}}\left(\boldsymbol{K}_{ii}\right)$ and $\boldsymbol{v}_t=\operatorname{col}_{i\in\mathbb{Z}_{1:M}}\left(\boldsymbol{v}_{i,t|t}^*\right)$. Moreover, note that through the relative triggering sets $\mathcal{E}_i^{\mathrm{r}}$, the global error $\boldsymbol{e}=\operatorname{col}_{i\in\mathbb{Z}_{1:M}}\left(\boldsymbol{e}_i\right)$ is bounded by

$$\boldsymbol{e}_t^{\mathrm{T}}\boldsymbol{S}\boldsymbol{e}_t\leq\sigma\boldsymbol{x}_t^{\mathrm{T}}\boldsymbol{S}\boldsymbol{x}_t\ \forall t\in\mathbb{N} \tag{5.30}$$

where $\boldsymbol{S}=\operatorname{diag}_{i\in\mathbb{Z}_{1:M}}\left(\boldsymbol{S}_i\right)$. For establishing attractivity of the origin, the following assumption is required.

Assumption 5.3.2. *The following matrix inequality is feasible for* $\boldsymbol{P}\succ\boldsymbol{0}$, $\boldsymbol{Q}_K\succ\boldsymbol{0}$, $\boldsymbol{N}\succ\boldsymbol{0}$ *and* $0<\delta<1$*:*

$$\begin{pmatrix}\boldsymbol{P}-\boldsymbol{F}^{\mathrm{T}}\boldsymbol{P}\boldsymbol{F}-\delta\boldsymbol{Q}_K-\sigma\boldsymbol{S} & \boldsymbol{F}^{\mathrm{T}}\boldsymbol{P}\boldsymbol{B}\boldsymbol{K}_{\mathrm{C}}\\ * & \boldsymbol{S}-\boldsymbol{K}_{\mathrm{C}}^{\mathrm{T}}\boldsymbol{B}^{\mathrm{T}}\boldsymbol{P}\boldsymbol{B}\boldsymbol{K}_{\mathrm{C}}-\boldsymbol{N}\end{pmatrix}\succeq\boldsymbol{0}. \tag{5.31}$$

The next two lemmas give an interpretation of the assumption and show that required parameters can always be selected such that it holds.

Lemma 5.3.3. *Consider the system* $\boldsymbol{x}_{t+1} = \boldsymbol{F}\boldsymbol{x}_t + \boldsymbol{B}\boldsymbol{K}_{\mathrm{C}}\boldsymbol{e}_t$ *and the function* $V^{\mathrm{f}}(\boldsymbol{x}) = \boldsymbol{x}^{\mathrm{T}}\boldsymbol{P}\boldsymbol{x}$*, assume that Assumption 5.3.2 holds and* $\boldsymbol{e}_t$ *is bounded according to* (5.30)*, then*

$$V^{\mathrm{f}}(\boldsymbol{x}_{t+1}) - V^{\mathrm{f}}(\boldsymbol{x}_t) \leq -\delta\boldsymbol{x}_t^T\boldsymbol{Q}_K\boldsymbol{x}_t - \boldsymbol{e}_t^T\boldsymbol{N}\boldsymbol{e}_t \tag{5.32}$$

$\forall t \in \mathbb{N}$ *holds.*

Lemma 5.3.4. *Suppose that* $\boldsymbol{F}$ *is stable, then there exists a* $\sigma > 0$ *such that Assumption 5.3.2 holds.*

The proofs can be found in Appendix B.3.

To obtain a suitable $\sigma > 0$ an LMI optimization problem can be solved. A suitable objective is to maximize σ to reduce the communication triggered by the relative trigger. Note that the solution of the optimization problem is dependent on δ. Obviously, the bigger δ (faster decay of the function $V^{\mathrm{f}}(\boldsymbol{x})$, see (5.32)), the smaller is σ (higher communication effort due to the relative trigger).

The results concerning convergence can now be stated.

Theorem 5.3.5. *Consider the closed-loop system* (5.29) *where the event-triggers and predictors are evaluated according to Algorithm 1 as well as the sequences* $\check{\boldsymbol{V}}_t$ *are determined by Algorithm 1, assume that* $\boldsymbol{x}_0 \in \mathcal{X}_N$ *and an initial sequence* $\check{\boldsymbol{V}}_0 \in \check{\mathcal{D}}_N(\boldsymbol{x}_0)$ *is given, then the state* $\boldsymbol{x}_t$ *converges asymptotically to the origin, i.e.,* $\lim_{t\to\infty}\|\boldsymbol{x}_t\| = 0$*. Furthermore,* $\lim_{t\to\infty}\|\boldsymbol{v}_t\| = 0$*,* $\lim_{t\to\infty}\|\boldsymbol{u}_t\| = 0$ *and* $\lim_{t\to\infty}\|\check{\boldsymbol{V}}_t\| = 0$*.*

The proof can be found in Appendix B.3.

5.4 Initialization

One way to determine initially feasible sequences $\check{\boldsymbol{V}}_0$ at time $t = 0$, i.e., $\check{\boldsymbol{V}}_0 \in \check{\mathcal{D}}_N(\boldsymbol{x}_0)$, is shown in Algorithm 2. With in it, the optimization problem

$$
\begin{aligned}
&\underset{\check{\check{\boldsymbol{V}}}_{i,0}}{\text{minimize}} \quad && J_{N,i}\left(\check{\check{\boldsymbol{V}}}_{i,0}\right) && \text{(5.33a)}\\
&\text{subject to} && \check{\check{\boldsymbol{x}}}_{i,N} \in \check{\check{\mathcal{X}}}_{\mathrm{f},i}, && \text{(5.33b)}\\
& && \check{\check{\boldsymbol{x}}}_{i,0} = \boldsymbol{x}_{i,0}, && \text{(5.33c)}\\
& && \text{and for } r \in \mathbb{Z}_{0:N-1}: && \text{(5.33d)}\\
& && \check{\check{\boldsymbol{x}}}_{i,r+1} = \boldsymbol{A}_{ii}\check{\check{\boldsymbol{x}}}_{i,r} + \boldsymbol{B}_{ii}\check{\check{\boldsymbol{u}}}_{i,r}, && \text{(5.33e)}\\
& && \check{\check{\boldsymbol{u}}}_{i,r} = \check{\check{\boldsymbol{v}}}_{i,r} + \boldsymbol{K}_{ii}\check{\check{\boldsymbol{x}}}_{i,r} && \text{(5.33f)}\\
& && \check{\check{\boldsymbol{u}}}_{i,r} \in \check{\check{\mathcal{U}}}_i^r, && \text{(5.33g)}\\
& && \check{\check{\boldsymbol{x}}}_{i,r} \in \check{\check{\mathcal{X}}}_i^r, && \text{(5.33h)}
\end{aligned}
$$

is solved. The constraints are determined by

$$
\begin{aligned}
&\check{\check{\mathcal{X}}}_i^r = \bar{\mathcal{X}}_i^r \ominus \check{\check{\mathcal{W}}}_i^r, && \forall r \in \mathbb{Z}_{0:N-1} && \text{(5.34)}\\
&\check{\check{\mathcal{U}}}_i^r = \bar{\mathcal{U}}_i^r \ominus \boldsymbol{K}_{ii}\check{\check{\mathcal{W}}}_i^r \ominus \bigoplus_{j\in\mathcal{N}_i} \boldsymbol{K}_{ij}\mathcal{X}_j, && \forall r \in \mathbb{Z}_{0:N-1} && \text{(5.35)}\\
&\check{\check{\mathcal{X}}}_{\mathrm{f},i} = (\bar{\mathcal{X}}_{\mathrm{f},i} \ominus \check{\check{\mathcal{W}}}_i^N) \cap \bar{\check{\mathcal{X}}}_{\mathrm{f},i}, && && \text{(5.36)}\\
&\check{\check{\mathcal{W}}}_i^r = \boldsymbol{F}_{ii}\check{\check{\mathcal{W}}}_i^{r-1} \oplus \bigoplus_{j\in\mathcal{N}_i} \boldsymbol{F}_{ij}\bar{\mathcal{X}}_j^{r-1} && \forall r \in \mathbb{Z}_{1:N} \text{ with } \check{\check{\mathcal{W}}}_i^0 = \{\boldsymbol{0}\}\,. && \text{(5.37)}
\end{aligned}
$$

The optimal solution is denoted as $\check{\check{\boldsymbol{V}}}_{i,0}^*$. The sequence is sent ot the neighbors and $\check{\boldsymbol{V}}_{j,0} = \check{\check{\boldsymbol{V}}}_{j,0}^*$ with $j \in \mathcal{N}_i \cup \{i\}$ are updated for all $i \in \mathbb{Z}_{1:M}$. Then, Algorithm 1 is started.

Lemma 5.4.1. *If optimization problem* (5.33) *is feasible and the sequences* $\check{\boldsymbol{V}}_{j,0}$ *are updated according to Algorithm 2 with* $j \in \mathcal{N}_i \cup \{i\}$ *for all* $i \in \mathbb{Z}_{1:M}$*, then a feasible solution to optimization problem* (5.18) *exists.*

Algorithm 2 Initialization for subsystem $\mathscr{S}_i$:

Measure $\boldsymbol{x}_{i,0}$

Obtain $\check{\check{\boldsymbol{V}}}_{i,0}^*$ as the solution of optimization problem (5.33)

Send $\check{\check{\boldsymbol{V}}}_{i,0}^*$ to neighbors

Receive $\check{\check{\boldsymbol{V}}}_{j,0}^*$ with $j \in \mathcal{N}_i$ and set $\check{\boldsymbol{V}}_{j,0} = \check{\check{\boldsymbol{V}}}_{j,0}^*$ for all $j \in \mathcal{N}_i \cup \{i\}$.

Go to Algorithm 1

The proof can be found in Appendix B.3.

For large prediction horizons the sets $\check{\check{\mathcal{X}}}_i^l$, $\check{\check{\mathcal{U}}}_i^l$ and $\check{\check{\mathcal{X}}}_i^{\mathrm{f}}$ can become empty. If the scheme is applicable, depends on the application at hand. Moreover, it should be noted that the proposed initialization is non-iterative. Another option is to use an iterative method to solve the initialization problem as suggested in [BFS14]. The algorithms are suited for the robust MPC method of [MSR05] and cannot be directly applied to the method presented here but the key ideas could be transfered. The resulting algorithms are expected to be less conservative but require more communication.

5.5 Illustrative Example

For illustration purpose, four coupled inverted pendulum mounted on carts that move in a horizontal direction as shown in Figure 5.2 are investigated. Each cart is driven by a motor that exerts a horizontal force $F_i, i \in \mathbb{Z}_{1:M}$ with $M = 4$ on the cart. Furthermore, the pendulum rods are mechanically connected by a spring with a spring coefficient $k = 15\,\mathrm{N/m}$. The pendulum rods are considered as mass-less with a length of $l = l_i = 0.75\,\mathrm{m}\ \forall i \in \mathbb{Z}_{1:M}$. The spring is attached at $L = 0.4\,\mathrm{m}$. In addition, each pendulum has a point mass $m = m_i = 0.3\,\mathrm{kg}\ \forall i \in \mathbb{Z}_{1:M}$ at the upper end. The cart mass is $M = M_i = 0.1\,\mathrm{kg}\ \forall i \in \mathbb{Z}_{1:M}$. The linearized state equation of each pendulum around the upright equilibrium can be described by

$$\begin{pmatrix} \dot{\phi}_i \\ \ddot{\phi}_i \end{pmatrix} = \begin{pmatrix} 0 & 1 \\ \frac{mgl+alkL-kL^2}{Mal^2} & 0 \end{pmatrix} \begin{pmatrix} \phi_i \\ \dot{\phi}_i \end{pmatrix} + \begin{pmatrix} 0 \\ -\frac{1}{Ml} \end{pmatrix} F_i + \sum_{j \in \mathcal{N}_i} \begin{pmatrix} 0 & 0 \\ \frac{kL^2-alkL}{Mal^2} & 0 \end{pmatrix} \begin{pmatrix} \phi_j \\ \dot{\phi}_j \end{pmatrix} \tag{5.38}$$

with $g = 9.81\ \mathrm{m/s^2}$ and $a = m/(M+m)$. The sets of neighboring subsystems are $\mathcal{N}_1 = \{2\}$, $\mathcal{N}_2 = \{1,3\}$, $\mathcal{N}_3 = \{2,4\}$ and $\mathcal{N}_4 = \{3\}$. The system is subject to the state

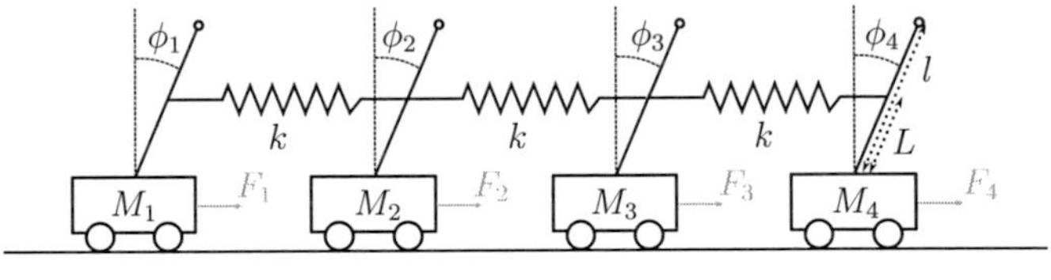

Figure 5.2: Considered set-up of simultaneous stabilization of coupled inverted pendulums mounted on carts

and input constraints $|\phi_i| \leq 0.1\,, |\dot{\phi}_i| \leq 2$ and $|F_i| \leq 5$ for all $i \in \mathbb{Z}_{1:4}$. In the following the discrete-time system which is obtained by discretizing the continuous-time plant (5.38) with Euler-forward and a fixed sampling interval of $h = 3.5\,\text{ms}$ is considered.

The matrix $\boldsymbol{P} = \operatorname{diag}_{i\in\mathbb{Z}_{1:M}}(\boldsymbol{P}_i)$ and the controller gain $\boldsymbol{K}$ are designed such that the Lyapunov inequality $\boldsymbol{F}^{\mathrm{T}}\boldsymbol{P}\boldsymbol{F} - \boldsymbol{P} \preceq -\boldsymbol{Q}_K$ holds with $\boldsymbol{Q}_K = \boldsymbol{Q} + \boldsymbol{K}^{\mathrm{T}}\boldsymbol{R}\boldsymbol{K}$ and the weighting matrices $\boldsymbol{Q}_i = \operatorname{diag}(1000, 1)$ and $\boldsymbol{R}_i = 1\ \forall i \in \mathbb{Z}_{1:M}$ by solving an LMI optimization problem. The weighting matrices are $\boldsymbol{M}_i = \boldsymbol{B}_{ii}^{\mathrm{T}}\boldsymbol{P}_i\boldsymbol{B}_{ii} + \boldsymbol{R}_i$. Moreover, the parameters in LMI (5.31) are chosen to $\delta = 0.9$ and $\sigma = 0.01$ so that Assumption 5.3.2 is satisfied. The prediction horizon is $N = 5$. The constraint on the change of the sequence is set to $\mathcal{V}_i = \{v_i \in \mathbb{R} \mid |v_i| \leq 0.5\}$. The absolute trigger sets are $\mathcal{E}_i^{\mathrm{a}} = \left\{[e_i^{[1]}, e_i^{[2]}]^{\mathrm{T}} \in \mathbb{R}^2 \mid |e_i^{[1]}| \leq 0.001, |e_i^{[2]}| \leq 0.05\right\}$. The tightened constraint sets are computed according to (5.20) and the terminal set are chosen to satisfy Assumption 5.2.2. The DMPC controller is initialized by computing initial trajectories $\check{\boldsymbol{V}}$ with Algorithm 2.

Figure 5.3 shows the state trajectory for the event-triggered DMPC scheme for an initial value chosen such that the system operates close to the constraints. The control law remains recursively feasible and the states converge to the origin satisfying the constraints at all time. At the beginning events are triggered due to the absolute trigger as the system is moving fast and the coupling causes a deviation between the actual state and the state obtained through the prediction model (5.7). When the state approaches the origin, the number of events reduces. Most of them are then caused by the relative threshold.

In Table 5.1 controllers designed with different absolute triggering sets of the form

$$\mathcal{E}_i^{\mathrm{a}} = \left\{[e_i^{[1]}, e_i^{[2]}]^{\mathrm{T}} \in \mathbb{R}^2 \mid |e_i^{[1]}| \leq e_i^{\mathrm{a}[1]}, |e_i^{[2]}| \leq e_i^{\mathrm{a}[2]}\right\} \tag{5.39}$$

are compared. The other data remains unchanged. Note that the controller with $e_i^{\mathrm{a}[1]} = e_i^{\mathrm{a}[2]} = 0$ communicates its state measurement to the neighboring subsystems in every time step, i.e., the communication is time-triggered. Each controller is simulated for 25 initial values for $N_{\mathrm{sim}} = 100$ steps. The initial values are chosen such that they lie

Table 5.1: Communication effort, cost and ROA for different e_i^{a}

$e_i^{\mathrm{a}[1]}$	$1 \cdot 10^{-3}$	$0.5 \cdot 10^{-3}$	0
$e_i^{\mathrm{a}[2]}$	$50 \cdot 10^{-3}$	$25 \cdot 10^{-3}$	0
c^{com}	21.58 %	32.85 %	100 %
J_{cl}	947.53	947.48	947.48
$A_{\mathrm{e}}/A_{\mathrm{e}=0}$	93.39 %	98.64 %	100 %

outside the terminal set but inside the ROA of the controller with the largest absolute trigger set. The quantities to be compared with each other are the closed-loop costs $J_{\text{cl}} = \sum_{t=0}^{N_{\text{sim}}} \boldsymbol{x}_t^{\text{T}} \boldsymbol{Q} \boldsymbol{x}_t + \boldsymbol{u}_t^{\text{T}} \boldsymbol{R} \boldsymbol{u}_t$, the number of communication events in each subsystem c^{com} in relation to the simulation time and the ratio of the area of the ROAs A_e to the one of the controller with time-triggered communication.

One can see that the performance and the ROA increases with decreasing threshold. The reduction in the ROA compared to a DMPC approach with communication in every time step is due to inclusion of the error caused by the event-triggered communication into the robust design in (5.13) and (5.17). At the same time the communication effort increases. Moreover, one can observe that the reduction of the size of the ROA and the performance deterioration of the event-triggered controllers is small compared to the time-triggered controller.

5.6 Summary

In this chapter a non-iterative DMPC scheme for constrained dynamically coupled subsystems with event-triggered communication of state measurements between the subsystems is presented.

The approach uses predictors for generating copies of the state values of the neighboring subsystems in case of no communication. Moreover, each subsystem has a predictor for the reconstruction of the copy of its own state in the other subsystems. This predictor is exploited for event-triggering. Events are triggered when the actual local state deviates too much from the predicted one. The applied threshold consist of an absolute and a relative triggering set. The absolute part is required in the design of the controller for ensuring recursive feasibility. With the help of the relative threshold, convergence of the state to the origin can be established. It is shown how the relative threshold can be designed using LMIs.

A consistency constraint is utilized in the controller to reduce the uncertainties the subsystems impose on each other through the coupling. The overall controller design is inspired by the robust MPC framework presented in Section 2.3. Furthermore, it is discussed how the algorithm can be initialized.

In simulation it is investigated how the absolute trigger parameters affect the performance, size of ROA and communication effort. Moreover, the results show that the communication effort can be reduced significantly compared to a time-triggered controller while the performance degradation is relatively small.

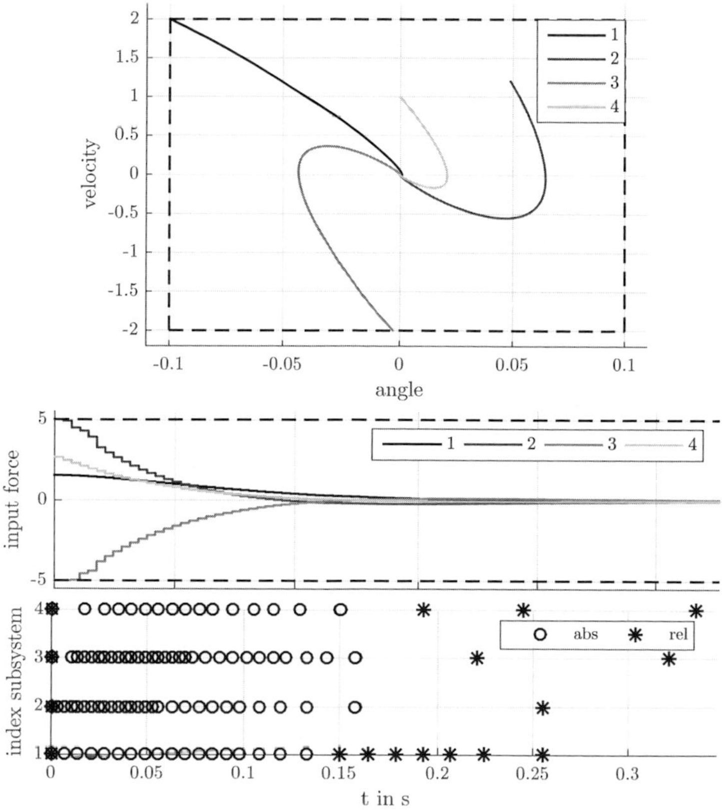

Figure 5.3: Simulation results for the coupled inverted pendulums for the initial value $\boldsymbol{x}_{1,0} = (-0.1, 2)^{\mathrm{T}}$, $\boldsymbol{x}_{2,0} = (0.05, 1.2)^{\mathrm{T}}$, $\boldsymbol{x}_{3,0} = (-0.003, -2)^{\mathrm{T}}$ and $\boldsymbol{x}_{4,0} = (0, 1)^{\mathrm{T}}$. The upper figure shows the state trajectories of the subsystems in solid lines. In the middle the input trajectories are depicted. The dashed black lines are the constraints. In the lower figure events in the subsystems are shown.

6 Event-triggered Cooperation in DMPC

In this chapter a decomposition-based DMPC scheme for systems consisting of dynamically coupled subsystems subject to constraints is presented where the controllers of the subsystems decide whether they solve the optimization problem with neighboring controllers in an event-triggered fashion. The chapter is based on [BL18a].

6.1 Problem Formulation

In this section, first, the considered set-up is introduced. Then, the nominal decomposition-based DMPC problem which can be solved with iterative DOAs is presented. Afterwards, the problem which is dealt with in this chapter is formulated and the main idea of the event-triggered cooperation approach is discussed.

6.1.1 Problem Set-up

The distributed control set-up depicted in Figure 3.2 is considered. The dynamics of each subsystem $\mathscr{S}_i$ are described by the LTI discrete-time model

$$\mathscr{S}_i : \boldsymbol{x}_{i,t+1} = \boldsymbol{A}_{ii}\boldsymbol{x}_{i,t} + \boldsymbol{B}_{ii}\boldsymbol{u}_{i,t} + \sum_{j \in \mathcal{N}_i} \boldsymbol{A}_{ij}\boldsymbol{x}_{j,t} + \boldsymbol{w}_{i,t} \tag{6.1}$$

for all $t \in \mathbb{N}$ which corresponds to the state-feedback case of (3.1) with coupling in the dynamics via the states. It is assumed that the inputs and states are subject to the local constraints

$$\boldsymbol{u}_{i,t} \in \mathcal{U}_i \text{ and } \boldsymbol{x}_{i,t} \in \mathcal{X}_i \tag{6.2}$$

for all $t \in \mathbb{N}$, respectively. $\mathcal{U}_i$ and $\mathcal{X}_i$ are convex compact sets which contain the origin in their interior.

Each subsystem can measure its local state $\boldsymbol{x}_i$ and can communicate with the neighboring subsystems over a communication network. Assumptions 3.3.1 and 3.3.2 hold, i.e., the components in the subsystems as well as all the subsystems are synchronized and the communication network is ideal. Note that similar to the case in Chapter 5, due to reasons of clarity the approach does not consider the output-feedback case. However, a combination with OB-MPC, e.g., as outlined in Chapter 4, is possible. The case of input coupling can also be considered by using the well-known delta input formulation (cf. [CJMZ16]).

Stacking up the dynamics (6.1), one obtains the global model

$$\mathscr{S}_i : \boldsymbol{x}_{t+1} = \boldsymbol{A}\boldsymbol{x}_t + \boldsymbol{B}\boldsymbol{u}_t + \boldsymbol{w}_t \tag{6.3}$$

for all $t \in \mathbb{N}$. For more information refer to Section 3.2. For the robust control design, the controller parametrization

$$\boldsymbol{u}_{i,t} = \boldsymbol{v}_{i,t} + \boldsymbol{K}_{ii}\boldsymbol{x}_{i,t} + \sum_{j\in\mathcal{N}_i} \boldsymbol{K}_{ij}\boldsymbol{x}_{j,t} \tag{6.4}$$

is utilized. $\boldsymbol{K}_{ij} \in \mathbb{R}^{m_i \times n_j}$ are feedback gains. The local dynamics (6.1) can be recast as

$$\boldsymbol{x}_{i,t+1} = \boldsymbol{F}_{ii}\boldsymbol{x}_{i,t} + \boldsymbol{B}_{ii}\boldsymbol{v}_{i,t} + \sum_{j\in\mathcal{N}_i} \boldsymbol{F}_{ij}\boldsymbol{x}_{j,t} + \boldsymbol{w}_{i,t} \tag{6.5}$$

where $\boldsymbol{F}_{ij} = \boldsymbol{A}_{ij} + \boldsymbol{B}_{ii}\boldsymbol{K}_{ij}$ $\forall j \in \mathcal{N}_i \cup \{i\}$, $i \in \mathbb{Z}_{1:M}$. The global input is given by $\boldsymbol{u}_t = \boldsymbol{v}_t + \boldsymbol{K}\boldsymbol{x}_t$ with $\boldsymbol{v} = \mathrm{col}_{i\in\mathbb{Z}_{1:M}}(\boldsymbol{v}_i)$ and $\boldsymbol{K} = (\boldsymbol{K}_{ij})$. Moreover, the matrix $\boldsymbol{F} = \boldsymbol{A} + \boldsymbol{B}\boldsymbol{K}$ is defined. For reasons as explained in Chapter 5, it has the same structure as $\boldsymbol{A}$.

Another more compact form of (6.1) which is used in the sequel is $\boldsymbol{x}_{i,t+1} = \boldsymbol{A}_{\mathcal{N}_i}\boldsymbol{x}_{\mathcal{N}_i,t} + \boldsymbol{B}_{ii}\boldsymbol{u}_{i,t}$ with the stacked state vector $\boldsymbol{x}_{\mathcal{N}_i} = \mathrm{col}_{j\in\mathcal{N}_i\cup\{i\}}(\boldsymbol{x}_j)$. $\boldsymbol{A}_{\mathcal{N}_i} \in \mathbb{R}^{n_i \times n_{\mathcal{N}_i}}$ is the corresponding system matrix. Equation (6.4) can also be written in the compact form $\boldsymbol{u}_{i,t} = \boldsymbol{v}_{i,t} + \boldsymbol{K}_{\mathcal{N}_i}\boldsymbol{x}_{\mathcal{N}_i,t}$.

6.1.2 Nominal DMPC Based on Distributed Optimization

Next, the decomposition-based DMPC problem that is solved online using DOAs is presented. For the sake of convenience, in this subsection the undistorted case is considered, i.e., in (6.1) the disturbance $\boldsymbol{w}_{i,t} = \boldsymbol{0}$ for all $i \in \mathbb{Z}_{1:M}$, $t \in \mathbb{N}$.

The online DMPC problem is of the form

$$\underset{\boldsymbol{V}_{i,t}}{\text{minimize}} \quad \sum_{i=1}^{M} J_{N,i}\left(\boldsymbol{V}_{i,t}\right) \tag{6.6a}$$

$$\text{subject to} \quad i \in \mathbb{Z}_{1:M}: \tag{6.6b}$$

$$\boldsymbol{x}_{i,t|t} = \boldsymbol{x}_{i,t} \tag{6.6c}$$

$$\boldsymbol{x}_{i,t+N|t} \in \mathcal{T}_i\left(\theta_{i,t}\right), \tag{6.6d}$$

$$\text{and } r \in \mathbb{Z}_{t:t+N-1}: \tag{6.6e}$$

$$\boldsymbol{x}_{i,r+1|t} = \sum_{j \in \mathcal{N}_i \cup \{i\}} \boldsymbol{A}_{ij}\boldsymbol{x}_{j,r|t} + \boldsymbol{B}_{ii}\boldsymbol{u}_{i,r|t} \tag{6.6f}$$

$$\boldsymbol{u}_{i,r|t} = \boldsymbol{v}_{i,r|t} + \sum_{j \in \mathcal{N}_i \cup \{i\}} \boldsymbol{K}_{ij}\boldsymbol{x}_{j,r|t} \in \mathcal{U}_i \tag{6.6g}$$

$$\boldsymbol{x}_{i,r|t} \in \mathcal{X}_i. \tag{6.6h}$$

$J_{N,i}\left(\boldsymbol{V}_{i,t}\right) = \sum_{r=t}^{t+N-1} l_i\left(\boldsymbol{v}_{i,r}\right)$ is the local cost function with $\boldsymbol{V}_{i,t} = \left[\boldsymbol{v}_{i,t|t}^{\mathrm{T}}, \dots, \boldsymbol{v}_{i,t+N-1|t}^{\mathrm{T}}\right]^{\mathrm{T}}$. The local stage cost is $l_i\left(\boldsymbol{v}_i\right) = \boldsymbol{v}_i^{\mathrm{T}}\boldsymbol{M}_i\boldsymbol{v}_i$. $\mathcal{T}_i\left(\theta_i\right)$ are local terminal constraint sets of the form $\mathcal{T}_i\left(\theta_i\right) = \left\{\boldsymbol{x}_i \in \mathbb{R}^{n_i} | V_i^{\mathrm{f}}\left(\boldsymbol{x}_i\right) \leq \theta_i\right\}$ where $\theta_i \in \mathbb{R}_{\geq 0}$ and the functions $V_i^{\mathrm{f}}\left(\boldsymbol{x}_i\right)$ are defined by $V_i^{\mathrm{f}}\left(\boldsymbol{x}_i\right) = \boldsymbol{x}_i^{\mathrm{T}}\boldsymbol{P}_i\boldsymbol{x}_i$ with $\boldsymbol{P}_i \in \mathbb{R}^{n_i \times n_i}$, $\boldsymbol{P}_i \succ \boldsymbol{0}$. The functions have to satisfy the following main assumption.

Assumption 6.1.1 ([CJMZ16]). *For all $i \in \mathbb{Z}_{1:M}$ there exist linear controller gains $\boldsymbol{K}_{\mathcal{N}_i}$, functions $\tau_i\left(\boldsymbol{x}_{\mathcal{N}_i}\right) = \boldsymbol{x}_{\mathcal{N}_i}^{\mathrm{T}}\boldsymbol{T}_{\mathcal{N}_i}\boldsymbol{x}_{\mathcal{N}_i}$ with $\boldsymbol{T}_{\mathcal{N}_i} \in \mathbb{R}^{n_{\mathcal{N}_i} \times n_{\mathcal{N}_i}}$, functions $q_i\left(\boldsymbol{x}_{\mathcal{N}_i}\right) = \boldsymbol{x}_{\mathcal{N}_i}^{\mathrm{T}}\boldsymbol{Q}_{\mathcal{N}_i}\boldsymbol{x}_{\mathcal{N}_i}$ with $\boldsymbol{Q}_{\mathcal{N}_i} \in \mathbb{R}^{n_{\mathcal{N}_i} \times n_{\mathcal{N}_i}}$, $\boldsymbol{Q}_{\mathcal{N}_i} \succ \boldsymbol{0}$ and a PI set $\mathcal{T} = \left\{\boldsymbol{x} \in \mathbb{R}^n | V^{\mathrm{f}}\left(\boldsymbol{x}\right) \leq \theta\right\}$ such that $\forall \boldsymbol{x} \in \mathcal{T}$:*

$$V_i^{\mathrm{f}}\left(\left(\boldsymbol{A}_{\mathcal{N}_i} + \boldsymbol{B}_{ii}\boldsymbol{K}_{\mathcal{N}_i}\right)\boldsymbol{x}_{\mathcal{N}_i}\right) \leq \tilde{V}_i^{\mathrm{f}}\left(\boldsymbol{x}_{\mathcal{N}_i}\right) \tag{6.7a}$$

$$\tilde{V}_i^{\mathrm{f}}\left(\boldsymbol{x}_{\mathcal{N}_i}\right) = V_i^{\mathrm{f}}\left(\boldsymbol{x}_i\right) - q_i\left(\boldsymbol{x}_{\mathcal{N}_i}\right) + \tau_i\left(\boldsymbol{x}_{\mathcal{N}_i}\right) \tag{6.7b}$$

$$\sum_{i=1}^{M} \tau_i\left(\boldsymbol{x}_{\mathcal{N}_i}\right) \leq 0. \tag{6.7c}$$

Notably, $\tilde{V}_i^{\mathrm{f}}\left(\boldsymbol{x}_{\mathcal{N}_i}\right)$ is introduced for notational convenience. From the assumption, it follows that the function $V^{\mathrm{f}}\left(\boldsymbol{x}\right) = \sum_{i=1}^{M} V_i^{\mathrm{f}}\left(\boldsymbol{x}_i\right) = \boldsymbol{x}^{\mathrm{T}}\boldsymbol{P}\boldsymbol{x}$ with $\boldsymbol{P} = \operatorname{diag}_{i \in \mathbb{Z}_{1:M}}\left(\boldsymbol{P}_i\right)$ is a structured Lyapunov function for the system $\boldsymbol{x}_{t+1} = \boldsymbol{F}\boldsymbol{x}_t$. The key idea of this choice is to allow a local increase of the Lyapunov-like functions V_i^{f} as long the global Lyapunov function V^{f} decreases. This follows from summing up (6.7a+b) over all subsystems and applying (6.7c) which yields $V^{\mathrm{f}}\left(\boldsymbol{F}\boldsymbol{x}\right) - V^{\mathrm{f}}\left(\boldsymbol{x}\right) \leq -q\left(\boldsymbol{x}\right)$ where $q\left(\boldsymbol{x}\right) = \sum_{i=1}^{M} q\left(\boldsymbol{x}_{\mathcal{N}_i}\right) = \boldsymbol{x}^{\mathrm{T}}\boldsymbol{Q}_K\boldsymbol{x}$. With the assumption it is also possible to design an update-law for the time-varying $\theta_{i,t}$ such that feasibility of (6.6) for all $t \in \mathbb{N}$ is maintained.

Optimization problem (6.6) has a decomposable structure which allows one to solve it efficiently using iterative DOAs. Algorithms that use the idea of dual decomposition are particularly suitable for this purpose. As they require communication in every iteration to coordinate the subsystems and a large number of iterations is usually required until a feasible and cost improving solution is found, the communication effort can be very high.

The objective is to design a scheme based on such DOAs where the controllers only solve the optimization problem cooperatively with their neighboring controllers when it is required, i.e., an event is triggered. Thereby, the communication effort is reduced as most data needs to be exchanged between the controllers when they cooperatively solve the problem. Simultaneously, recursive feasibility and asymptotic stability of a set around the origin have to be established.

Due to the disturbance and the error caused by the event-triggered cooperation, a robust controller design is required. At the same time, the optimization problem needs to maintain its decomposable structure, so that it can be solved efficiently by the DOAs. Here, the scheme adopts the concept of time-varying terminal sets as proposed in [CJMZ16] for nominal DMPC and extends it with the ideas from [CRZ01] and [LAC02] to account for the uncertainties and the time-varying cooperation structure which is caused by the event-triggering.

In the next section the event-triggered cooperation approach is presented. In Section 6.3, its main properties are analyzed and the main assumptions are stated. In Section 6.4 an illustrative example is given. Section 6.5 summarizes the method.

Remark 6.1.2. *The matrices* $\boldsymbol{P}_i$ *and* $\boldsymbol{T}_{\mathcal{N}_i}$ *as well as the feedback gains* $\boldsymbol{K}_{\mathcal{N}_i}$ *can be determined by solving an LMI problem offline. For this purpose, DOAs such as ADMM can be used and thus the approach is also applicable to large-scale systems. Due to the structural assumptions on* $\boldsymbol{K}_{\mathcal{N}_i}$ *and* V_i^{f} *conservatism is introduced. For more information refer to [CJMZ16].*

6.2 Robust DMPC with Event-triggered Cooperation

To enable the event-triggered cooperative optimization, each controller has exchanged its predicted state sequences with its neighboring controllers in a previous time step. Each controller then solves the optimization problem using the previously exchanged state sequences. If the sequence obtained in the iterations of the algorithm differs too much from the previously exchanged state sequences, the local controller sends a request to its neighboring controllers in order to solve the optimization problem cooperatively.

Due to the event-triggering the cooperation structure is time-varying. In this section, it is first shown how optimization problem (6.6) can be adapted to include the time-varying cooperation structure. Afterwards an algorithm which determines the cooperation structure is presented. Note that from now on the robust problem is considered again, i.e., the disturbance $\boldsymbol{w}_{i,t} \in \mathcal{W}_i$ for all $i \in \mathbb{Z}_{1:M}$ and $t \in \mathbb{N}$.

The DMPC optimization problem to be solved is

$$
\begin{aligned}
\underset{\boldsymbol{V}_{i,t}}{\text{minimize}} \quad & \sum_{i=1}^{M} J_{N,i}\left(\boldsymbol{V}_{i,t}\right) && \text{(6.8a)}\\
\text{subject to} \quad & \text{for all } i \in \mathbb{Z}_{1:M}: \\
& \bar{\boldsymbol{x}}_{i,t|t} = \boldsymbol{x}_{i,t}, && \text{(6.8b)}\\
& \bar{\boldsymbol{x}}_{i,t+N|t} \in \mathcal{X}_i^{\mathrm{f}}\left(\alpha_{i,t}\right) && \text{(6.8c)}\\
& \text{for all } r \in \mathbb{Z}_{0:N-1}: \\
& \bar{\boldsymbol{x}}_{i,t+r+1|t} = \boldsymbol{A}_{ii}\bar{\boldsymbol{x}}_{i,t+r|t} + \boldsymbol{B}_{ii}\bar{\boldsymbol{u}}_{i,t+r|t} + \boldsymbol{w}^{\mathrm{c}}_{i,t+r|t} && \text{(6.8d)}\\
& \boldsymbol{w}^{\mathrm{c}}_{i,t+r|t} = \sum_{j\in\mathcal{C}_{i,t}} \boldsymbol{A}_{ij}\bar{\boldsymbol{x}}_{j,t+r|t} + \sum_{j\in\mathcal{N}_i\setminus\mathcal{C}_{i,t}} \boldsymbol{A}_{ij}\check{\boldsymbol{x}}^{[i]}_{j,t+r|t} && \text{(6.8e)}\\
& \bar{\boldsymbol{u}}_{i,t+r|t} = \boldsymbol{v}_{i,t+r|t} + \sum_{j\in\mathcal{C}_{i,t}\cup\{i\}} \boldsymbol{K}_{ij}\bar{\boldsymbol{x}}_{j,t+r|t} + \sum_{j\in\mathcal{N}_i\setminus\mathcal{C}_{i,t}} \boldsymbol{K}_{ij}\check{\boldsymbol{x}}^{[i]}_{j,t+r|t} && \text{(6.8f)}\\
& \bar{\boldsymbol{u}}_{i,t+r|t} \in \bar{\mathcal{U}}_i^{r} && \text{(6.8g)}\\
& \bar{\boldsymbol{x}}_{i,t+r|t} \in \bar{\mathcal{X}}_i^{r}. && \text{(6.8h)}
\end{aligned}
$$

Compared to problem (6.6), the prediction of the future states and inputs in (6.8d-f) incorporates $\check{\boldsymbol{x}}^{[i]}_{j,r}$ which is subsystem $\mathscr{S}_i$'s copy of the predicted state for time $r \in \mathbb{Z}_{t:t+N-1}$ of subsystem $\mathscr{S}_j$ which has been sent in a previous time step. $\mathcal{C}_{i,t} \subseteq \mathcal{N}_i$ is the set of subsystems with which the controller of subsystem $\mathscr{S}_i$ cooperates with at time t. The global cooperation structure at time t is $\mathcal{C}_t = \{\mathcal{C}_{1,t}, \ldots, \mathcal{C}_{M,t}\}$. An example for the event-triggered cooperation structure is given in Figure 6.1. $\bar{\mathcal{X}}_i^r$ and $\bar{\mathcal{U}}_i^r$ are local tightened state and input constraints, respectively. $\mathcal{X}_i^{\mathrm{f}}(\alpha_i) = \left\{\boldsymbol{x}_i \in \mathbb{R}^{n_i} \middle| V_i^{\mathrm{f}}(\boldsymbol{x}_i) \leq \alpha_i\right\}$ are the

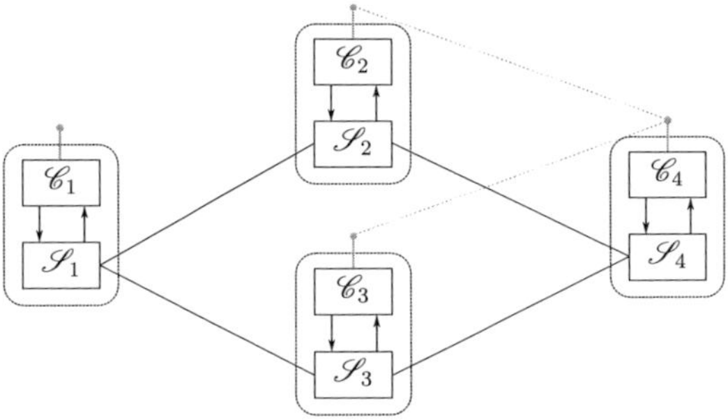

Figure 6.1: The figure illustrates the cooperation structure for an example of a system with $M = 4$ subsystems. The black solid links stand for the physical coupling and the communication links between the subsystems. The dashed gray line shows the cooperative optimization of controllers of the subsystems $\mathscr{S}_2$, $\mathscr{S}_3$ and $\mathscr{S}_4$, i.e., $\mathcal{C}_2 = \{4\}$, $\mathcal{C}_3 = \{4\}$ and $\mathcal{C}_4 = \{2,3\}$ as well as $\mathcal{C}_1 = \emptyset$.

local terminal sets which depend on $\alpha_i \in \mathbb{R}_{>0}$. The function V_i^{f} is designed as described in Subsection 6.1.2. The sets $\mathcal{X}_i^{\mathrm{f}}$ have to satisfy additional assumptions presented in Subsection 6.3.1 to ensure recursive feasibility.

Local event-triggers force the decision to solve the optimization problem cooperatively with neighboring controllers when the deviation of the last transmitted state sequences from the state sequences in the current iteration (obtained from optimization as shown later) leaves a set $\mathcal{E}_i^{\mathrm{x}}$:

$$\begin{aligned}&\exists \bar{\boldsymbol{x}}_{i,r|t}, \check{\boldsymbol{x}}_{i,r|t}^{[j]} \text{ for any } r \in \mathbb{Z}_{t:t+N-1}, \text{ s.t. } \bar{\boldsymbol{x}}_{i,r|t} - \check{\boldsymbol{x}}_{i,r|t}^{[j]} \notin \mathcal{E}_i^{\mathrm{x}} \\ &\Rightarrow j \in \mathcal{C}_{i,t} \wedge i \in \mathcal{C}_{j,t}.\end{aligned} \tag{6.9}$$

The set $\mathcal{E}_i^{\mathrm{x}}$ is convex, compact and $\mathbf{0} \in \mathcal{E}_i^{\mathrm{x}}$. It describes the allowed deviation between the sequences. Since it is independent of the current state, it is comparable to the absolute trigger set in Chapter 4 and 5. In general, the bigger the size of $\mathcal{E}_i^{\mathrm{x}}$ is chosen, the less likely it is that the controller of subsystem $\mathscr{S}_i$ will cooperate with its neighboring controllers. However, to ensure recursive feasibility, the size of $\mathcal{E}_i^{\mathrm{x}}$ has to be bounded as shown in Section 6.3.1. From (6.9) it follows that, for the subsystems with which no joint cooperative optimization has been triggered in the controller of subsystem $\mathscr{S}_i$, the deviation of the sequences is bounded, i.e.,

$$\bar{\boldsymbol{x}}_{j,r|t} - \check{\boldsymbol{x}}_{j,r|t}^{[i]} \in \mathcal{E}_j^{\mathrm{x}} \quad \forall j \in \mathcal{N}_i \setminus \mathcal{C}_{i,t}, \forall r \in \mathbb{Z}_{t:t+N-1}. \tag{6.10}$$

Algorithm 3 DMPC algorithm in every subsystem $\mathscr{S}_i$

Require: $\boldsymbol{x}_{i,t}$, $\boldsymbol{V}_{i,t}^{\text{shift}}$, $\breve{\boldsymbol{x}}_{j,t+r|t}^{[i]}$ $\forall j \in \mathcal{N}_i$, $r \in \mathbb{Z}_{0:N-1}$, $\delta_{\text{Xf},t}$

1: Send $\boldsymbol{x}_{i,t}$ to subsystems $\mathscr{S}_j$ with $j \in \mathcal{N}_i$
2: Receive $\boldsymbol{x}_{j,t}$ and set $\breve{\boldsymbol{x}}_{j,t|t}^{[i]} := \boldsymbol{x}_{j,t}$ $\forall j \in \mathcal{N}_i$
3: Set $\mathcal{C}_{i,t}^0 := \emptyset \quad \forall i \in \mathbb{Z}_{1:M}$, $k := 0$
4: **if** $\boldsymbol{x}_t \in \mathcal{X}_{\text{Glo}}^{\text{f}}$ or $\delta_{\text{Xf},t} = 1$ **then**
5:     Set $\boldsymbol{V}_{i,t} := \boldsymbol{0}$ $\forall i \in \mathbb{Z}_{1:M}$, and set $\delta_{\text{Xf},t+1} = 1$
6:     Go to step 27
7: **end if**
8: **while** $k \leq k_{\max}$ **do**
9:     Do Δk iteration of DOA to solve (6.8) with $\mathcal{C}_{i,t}^k$; Set $k := k + \Delta k$
10:     **if** $\exists \bar{\boldsymbol{x}}_{i,t+r}^k - \breve{\boldsymbol{x}}_{i,t+r|t}^{[j]} \notin \mathcal{E}_i^{\text{x}} \quad \forall r \in \mathbb{Z}_{0:N-1}, j \in \mathcal{N}_i$ **then**
11:         $\mathcal{C}_{i,t}^k := \mathcal{C}_{i,t}^{k-\Delta k} \cup \{j\}$, $\mathcal{C}_{j,t}^k := \mathcal{C}_{j,t}^{k-\Delta k} \cup \{i\}$
12:     **end if**
13:     **if** $\mathcal{C}_{i,t}^k = \mathcal{C}_{i,t}^{k-\Delta k}$ **then**
14:         Check feasibility and compute $J_{N,i}\left(\boldsymbol{V}_{i,t}^k\right)$
15:         **if** feasible and $J_N\left(\boldsymbol{V}_t^k\right) \leq J_N\left(\boldsymbol{V}_t^{\text{Shift}}\right)$ **then**
16:             Set $\boldsymbol{V}_{i,t} = \boldsymbol{V}_{i,t}^k$ $\forall i \in \mathbb{Z}_{1:M}$, $\mathcal{C}_{i,t} = \mathcal{C}_{i,t}^k$
17:             Go to step 24
18:         **end if**
19:     **end if**
20:     **if** $k = k_{\max}$ **then**
21:         Set $\boldsymbol{V}_{i,t} = \boldsymbol{V}_{i,t}^{\text{shift}}$ $\forall i \in \mathbb{Z}_{1:M}$
22:     **end if**
23: **end while**
24: Update $\breve{\boldsymbol{x}}_{j,t+r|t+1}^{[i]} = \bar{\boldsymbol{x}}_{j,t+r|t}$ $\forall r \in \mathbb{Z}_{1:N}$ $\forall j \in \mathcal{C}_{i,t}$. Update $\breve{\boldsymbol{x}}_{j,t+r|t+1}^{[i]} = \breve{\boldsymbol{x}}_{j,t+r|t}^{[i]}$ $\forall r \in \mathbb{Z}_{1:N-1}$ $\forall j \in \mathcal{N}_i \setminus \mathcal{C}_{i,t}$.
25: Send $\bar{\boldsymbol{x}}_{i,t+N|t}$ to neighboring subsystems $\mathscr{S}_j$ with $\forall j \in \mathcal{N}_i \setminus \mathcal{C}_{i,t}$. Update $\breve{\boldsymbol{x}}_{j,t+N|t+1}^{[i]} = \bar{\boldsymbol{x}}_{j,t+N|t}$ $\forall j \in \mathcal{N}_i \setminus \mathcal{C}_{i,t}$.
26: Update terminal set $\mathcal{X}_i^{\text{f}}$ according to (6.21)
27: Build $\boldsymbol{V}_{i,t+1}^{\text{shift}} = \left[\boldsymbol{v}_{i,t+1|t}^{\text{T}}, \boldsymbol{v}_{i,t+2|t}^{\text{T}}, \ldots, \boldsymbol{v}_{i,t+N-1|t}^{\text{T}}, \boldsymbol{0}\right]^{\text{T}}$
28: Apply $\boldsymbol{u}_{i,t} = \boldsymbol{v}_{i,t|t} + \boldsymbol{K}_{ii}\boldsymbol{x}_{i,t} + \sum_{j \in \mathcal{N}_i} \boldsymbol{K}_{ij}\boldsymbol{x}_{j,t}$ to plant

Next Algorithm 3 is discussed. It is used to find a cooperation structure $\mathcal{C}_t$, sequences $\boldsymbol{V}_{i,t}$ and the input $\boldsymbol{u}_t$ in each time step. k is the iteration counter of the algorithm. The algorithms searches for a solution with a minimum number of controllers cooperating. It requires the measurement $\boldsymbol{x}_{i,t}$, exchanged state sequences $\breve{\boldsymbol{x}}^{[i]}_{j,r|t}$ from the neighbors $j \in \mathcal{N}_i$, a shifted sequence $\boldsymbol{V}^{\text{shift}}_{i,t}$ and the variable $\delta_{\mathbb{X}\text{f}} \in \mathbb{Z}_{0:1}$. The exchanged state sequences $\breve{\boldsymbol{x}}^{[i]}_{j,r|t}$ as well as the shifted sequence $\boldsymbol{V}^{\text{shift}}_{i,t}$ are known from the previous time step. For the initialization at time $t = 0$ they can be obtained from the DOA as discussed in Section 6.3. $\delta_{\mathbb{X}\text{f},t} \in \mathbb{Z}_{0:1}$ indicates if $\boldsymbol{x}_t$ has entered the terminal set yet. For $t = 0$ it is initialized with $\delta_{\mathbb{X}\text{f},0} := 0$.

At the beginning, the subsystems exchange the measurements (step 1) and update the exchanged sequences with it (step 2). The cooperative optimization algorithm is started if the state has not entered the set $\mathcal{X}^{\text{f}}_{\text{Glo}}$ (queried in step (4)). The key idea of the cooperative optimization is the following: For a fixed cooperation structure, a suitable iterative DOA using N2N communication is used for solving problem (6.8) (step 9). For the subsystems $\mathscr{S}_j$ $j \in \mathcal{N}_i \setminus \mathcal{C}_{i,t}$ with which the controller of subsystem $\mathscr{S}_i$ is not cooperating at time t, a state sequence of the neighboring subsystems from time $t-1$ is known. The deviation of the state sequence in the current iteration from the last transmitted sequence is used as a triggering condition according to (6.9) (step 10). If (6.9) is violated for some neighboring subsystem $\mathscr{S}_j$, the controller of subsystem $\mathscr{S}_i$ sends a request to the controller of subsystem $\mathscr{S}_j$ and includes the neighboring subsystem into the cooperation structure and vice versa (step 11). To obtain a cooperation structure where only few controllers cooperate, the controllers start to solve the problem not cooperating at the beginning (i.e. $\mathcal{C}^0_{i,t} := \emptyset$, step 3). The procedure is repeated until no further events are triggered and the distributed optimization algorithm has found a feasible solution with lower cost than the shifted sequence $\boldsymbol{V}^{\text{shift}}_t$ (queried in step 15) where $J_N(\boldsymbol{V}_t) = \sum_{i=1}^{M} J_{N,i}(\boldsymbol{V}_{i,t})$ is called the global cost function. $\boldsymbol{V}_t$ is the collection of the sequences $\boldsymbol{V}_{i,t}$ for all $i \in \mathbb{Z}_{1:M}$. The shifted sequence $\boldsymbol{V}^{\text{shift}}_{i,t+1}$ for the next time step is generated from the current solution $\boldsymbol{V}_{i,t}$ in step 27 by taking the predicted $\boldsymbol{v}_{i,r|t}$ for $r \in \mathbb{Z}_{1:N-1}$ and extending it with the zero vector. It is a feasible sequence for the optimization problem at time $t+1$ and ensures a non-increasing cost from time t to $t+1$ as shown in the next section. When for instance dual-based DOAs are used, the feasibility and non-increasing cost property can get lost during the iterations. This is why a maximum number of iteration $k_{\max}$ is set. When it is reached, the shifted sequence is used (step 21). Note that the DOA is executed for $\Delta k \in \mathbb{N}$ iterations so that the break conditions do not need to be checked every iteration. The maximum number of iterations $k_{\max}$ must be a multiple of Δk, i.e., $k_{\max} = l \cdot \Delta k$ with $l \in \mathbb{N}$. After the optimization terminates, the exchanged sequences are updated for the next step. The terminal state of the predicted sequence is exchanged with the neighboring subsystems $\mathscr{S}_j$ with $j \in \mathcal{N}_i \setminus \mathcal{C}_{i,t}$, i.e., the neighboring subsystems which have not cooperatively solved the optimization problem with the controller of subsystem $\mathscr{S}_i$ (step 25). The terminal set is updated using the update law given in Section 6.3.1 (step 26). The first input is applied to the system (step 28).

Remark 6.2.1. *The algorithm is not communication-free even in the case when the controllers solve the optimization problem without any cooperation. The algorithm requires information exchange with other systems in the steps 1, 4, 9, 11, 14-15 and 24-25. In a fully cooperating algorithm, i.e.,* $\mathcal{C}_{i,t} = \mathcal{N}_i$ *for all* $i \in \mathbb{Z}_{1:M}$*, steps 11, 24 and 25 would not be required. However, the communication effort for these steps is relatively low compared to the other steps. Note that in all steps N2N communication is required. For checking the conditions in the steps 4 and 15 Algorithm 3 from [BFS14] can be applied.*

6.3 Analysis

In this section the algorithm is analyzed and the robust control design is discussed in more detail. Therefore, some additional definitions are required. The global input applied to (6.3) is obtained by stacking up the local inputs, i.e.,

$$\boldsymbol{u}_t = \mathrm{col}_{i\in\mathbb{Z}_{1:M}}\left(\boldsymbol{u}_{i,t}\right). \tag{6.11}$$

The collective closed-loop dynamics of (6.3) under control law (6.11) can be recast in the form

$$\begin{aligned}\boldsymbol{x}_{t+1} &= \boldsymbol{F}\boldsymbol{x}_t + \boldsymbol{B}\boldsymbol{v}_t + \boldsymbol{w}_t\\ \boldsymbol{V}_{t+1} &= \boldsymbol{h}\left(\boldsymbol{V}_t, \boldsymbol{x}_t\right)\end{aligned} \tag{6.12}$$

where the Algorithm 3 is collected in the mapping $\boldsymbol{h}(\cdot)$. The set of states where problem (6.8) is feasible is called $\mathcal{X}_N$ and $\boldsymbol{v}_t = \mathrm{col}_{i\in\mathbb{Z}_{1:M}}\left(\boldsymbol{v}_{i,t|t}\right)$.

6.3.1 Feasibility

As the cooperation structure $\mathcal{C}$ is determined by Algorithm 3, it can change arbitrarily between two time steps. To prevent loss of feasibility of optimization problem (6.8), a constraint tightening approach is applied which considers this change. The next lemma shows how the deviation of the state sequence at time t and $t+1$ can be bounded.

Lemma 6.3.1. *Consider the system (6.8d+e) and assume that*

$$\boldsymbol{V}_{i,t} = \left[\boldsymbol{v}_{i,t|t}^{\mathrm{T}}, \ldots, \boldsymbol{v}_{i,t+N-1|t}^{\mathrm{T}}\right]^{\mathrm{T}}$$

are feasible sequences $\forall i \in \mathbb{Z}_{1:M}$ *and the triggering conditions* (6.10) *are satisfied. For an arbitrary change of the cooperation structure from* $\mathcal{C}_t$ *to* $\mathcal{C}_{t+1}$ *and the corresponding state sequences exchanged according to Algorithm 3, the predicted state sequences starting from* $\boldsymbol{x}_{i,t+1}$ *constructed with the sequences* $\boldsymbol{V}_{i,t+1} = \left[\boldsymbol{v}_{i,t+1|t}^{\mathrm{T}}, \dots, \boldsymbol{v}_{i,t+N-1|t}^{\mathrm{T}}, \boldsymbol{0}\right]^{\mathrm{T}}$ *satisfy*

$$\bar{\boldsymbol{x}}_{i,t+r|t+1} - \bar{\boldsymbol{x}}_{i,t+r|t} \in \mathcal{H}_i^r \qquad \forall r \in \mathbb{Z}_{1:N} \tag{6.13}$$

$$\mathcal{H}_i^r = \mathcal{J}_i \oplus \bigoplus_{j \in \mathcal{N}_i \cup \{i\}} \boldsymbol{F}_{ij} \mathcal{H}_j^{r-1} \qquad \forall r \in \mathbb{Z}_{2:N} \tag{6.14}$$

$$\mathcal{H}_i^1 = \mathcal{W}_i \text{ and } \mathcal{J}_i = \bigoplus_{j \in \mathcal{N}_i} \boldsymbol{F}_{ij} \mathcal{E}_j^x \qquad i \in \mathbb{Z}_{1:M}. \tag{6.15}$$

The proof can be found in Appendix B.4.

For the global state, it follows that $\bar{\boldsymbol{x}}_{t+r|t+1} - \bar{\boldsymbol{x}}_{t+r|t} \in \mathcal{H}^r$ where $\mathcal{H}^r = \prod_{i=1}^M \mathcal{H}_i^r$ for all $r \in \mathbb{Z}_{1:N}$. To ensure recursive feasibility, the constraints are chosen such that for all $i \in \mathbb{Z}_{1:M}$

$$\bar{\mathcal{X}}_i^r = \bar{\mathcal{X}}_i^{r-1} \ominus \mathcal{H}_i^r \qquad \forall r = \mathbb{Z}_{1:N-1},\ \bar{\mathcal{X}}_i^0 = \mathcal{X}_i \tag{6.16}$$

$$\bar{\mathcal{U}}_i^r = \bar{\mathcal{U}}_i^{r-1} \ominus \bigoplus_{j \in \mathcal{N}_i \cup \{i\}} \boldsymbol{K}_{ij} \mathcal{H}_j^r \ominus \bigoplus_{j \in \mathcal{N}_i} \boldsymbol{K}_{ij} \mathcal{E}_j^{\mathrm{x}} \qquad \forall r = \mathbb{Z}_{1:N-1},\ \bar{\mathcal{U}}_i^0 = \mathcal{U}_i. \tag{6.17}$$

The global sets are obtained by $\bar{\mathcal{X}}^r = \prod_{i=1}^M \bar{\mathcal{X}}_i^r$ and $\bar{\mathcal{U}}^r = \prod_{i=1}^M \bar{\mathcal{U}}_i^r$ $\forall r \in \mathbb{Z}_{0:N-1}$. Next, some sets are introduced which are required for the design of the sets $\mathcal{X}_i^{\mathrm{f}}$.

Assumption 6.3.2. *There exist sets*

$$\mathcal{Y} = \left\{\boldsymbol{x} \in \mathbb{R}^n | V^{\mathrm{f}}(\boldsymbol{x}) \leq \beta\right\} \tag{6.18}$$

$$\text{and } \mathcal{X}_{\mathrm{Glo}}^{\mathrm{f}} = \left\{\boldsymbol{x} \in \mathbb{R}^n | V^{\mathrm{f}}(\boldsymbol{x}) \leq \alpha\right\} \tag{6.19}$$

with $\alpha, \beta \in \mathbb{R}_{>0}$ *such that*

1. $\mathcal{Y} \subseteq \bar{\mathcal{X}}^{N-1}$, $\boldsymbol{K}\mathcal{Y} \subseteq \bar{\mathcal{U}}^{N-1}$
2. $\boldsymbol{F}\mathcal{Y} \subseteq \mathcal{X}_{\mathrm{Glo}}^{\mathrm{f}}$
3. $\mathcal{X}_{\mathrm{Glo}}^{\mathrm{f}} \oplus \mathcal{H}^N \subseteq \mathcal{Y}$

hold.

If Assumption 6.1.1 is satisfied and the constraint sets are polytopes, then the set $\mathcal{Y}$ can be found by determining the maximum level set of the function V^{f} such that Assumption 6.3.2 item 1 holds by solving an Linear Program (LP). The second assumption requires that $\mathcal{Y}$ is a contractive set. The last part requires that the effect of the disturbance and non-cooperation (included in $\mathcal{H}^N$) is limited. The condition can be checked in the following way. Let Δ^{f} be defined as $\Delta^{\mathrm{f}} = \max_{\boldsymbol{x}\in\mathcal{X}^{\mathrm{f}}_{\mathrm{Glo}},\boldsymbol{y}\in\mathcal{Y}} \left(V^{\mathrm{f}}(\boldsymbol{y}) - V^{\mathrm{f}}(\boldsymbol{x})\right)$. Δ^{f} exist as V^{f} is a continuous functions as well as $\mathcal{X}^{\mathrm{f}}_{\mathrm{Glo}}$ and $\mathcal{Y}$ are bounded. If α is chosen as $\alpha = \beta\left(1 - \frac{\lambda^{\mathrm{m}}(\boldsymbol{Q}_K)}{\lambda^{\mathrm{M}}(\boldsymbol{P})}\right)$ and $\Delta^{\mathrm{f}} \leq \frac{\lambda^{\mathrm{m}}(\boldsymbol{Q}_K)}{\lambda^{\mathrm{M}}(\boldsymbol{P})}\beta$, then Assumption 6.3.2 item 2 and 3 hold. In total, Assumption 6.3.2 says that $\mathcal{X}^{\mathrm{f}}_{\mathrm{Glo}}$ is an RPI set which is a standard requirement for many robust MPC approaches (cf. Section 2.3). However, due to the distributed nature of the terminal set in (6.8) a further assumption is required. Therefore, let $\tilde{\Delta}^{\mathrm{f}}_i$ be defined as

$$\tilde{\Delta}^{\mathrm{f}}_i = \max_{\boldsymbol{x}\in\mathcal{X}^{\mathrm{f}}_{\mathrm{Glo}},\boldsymbol{y}\in\mathcal{Y}} \left(\tilde{V}^{\mathrm{f}}_i(\boldsymbol{W}_{\mathcal{N}_i}\boldsymbol{y}) - \tilde{V}^{\mathrm{f}}_i(\boldsymbol{W}_{\mathcal{N}_i}\boldsymbol{x})\right), \tag{6.20}$$

$\forall i \in \mathbb{Z}_{1:M}$ and where $\boldsymbol{W}_{\mathcal{N}_i} \in \mathbb{R}^{n_{\mathcal{N}_i}\times n}$ is a transformation matrix that maps the global state $\boldsymbol{x}$ on the state $\boldsymbol{x}_{\mathcal{N}_i}$. Note that $\tilde{\Delta}^{\mathrm{f}}_i$ exists as $\tilde{V}^{\mathrm{f}}_i$ are continuous functions as well as $\mathcal{X}^{\mathrm{f}}_{\mathrm{Glo}}$ and $\mathcal{Y}$ are bounded. Moreover, let $\tilde{\Delta}^{\mathrm{f}} = \sum_{i=1}^{M}\tilde{\Delta}^{\mathrm{f}}_i$.

Assumption 6.3.3. *$\tilde{\Delta}^{\mathrm{f}}$ is bounded by $\tilde{\Delta}^{\mathrm{f}} \leq \frac{\lambda^{\mathrm{m}}(\boldsymbol{Q}_K)}{\lambda^{\mathrm{M}}(\boldsymbol{P})}\beta$.*

The assumption implies that $\forall \boldsymbol{x} \in \mathcal{X}^{\mathrm{f}}_{\mathrm{Glo}}$: $V^{\mathrm{f}}(\boldsymbol{x}) - q(\boldsymbol{x}) + \tilde{\Delta}^{\mathrm{f}} \leq \alpha$. This is exploited in the proof of the following lemma. The size of the sets $\mathcal{H}^{\tau}_i$ as well as the values $\tilde{\Delta}^{\mathrm{f}}$ and Δ^{f} can be influenced by the choice of the sets $\mathcal{E}^{\mathrm{x}}_i$. The smaller the size of $\mathcal{E}^{\mathrm{x}}_i$ the more likely it is that Assumptions 6.3.2 and 6.3.3 can be fulfilled. On the contrary, it is expected that the number of time instants where the controllers cooperate increases. Next, the update law for the terminal set is introduced.

Lemma 6.3.4. *Consider the closed-loop system* (6.12)*. Assume that $\boldsymbol{x}_0 \in \mathcal{X}_N$, assumptions 6.3.2 and 6.3.3 hold, $\sum_{i=1}^{M}\alpha_{i,t} \leq \alpha$ and a feasible sequence $\boldsymbol{V}_t$ for optimization problem* (6.8) *for a given cooperation structure $\mathcal{C}_t$ is given. If the local terminal sets are updated according to*

$$\alpha_{i,t+1} := \tilde{V}^{\mathrm{f}}_i\left(\bar{\boldsymbol{x}}_{\mathcal{N}_i,t+N|t}\right) + \tilde{\Delta}^{\mathrm{f}}_i \quad \forall t \in \mathbb{N}, \tag{6.21}$$

then a feasible sequence $\boldsymbol{V}_{t+1}$ for an arbitrary cooperation structure $\mathcal{C}_{t+1}$ for optimization problem (6.8) *exists.*

The proof can be found in Appendix B.4.

For the cost of the optimization problem, the following important property holds.

Lemma 6.3.5. *Consider the closed-loop system* (6.12) *and assume $\boldsymbol{x}_0 \in \mathcal{X}_N$ as well as initially feasible sequences $\boldsymbol{V}_t$ are given, then for the global cost J_N*

$$J_N(\boldsymbol{V}_{t+1}) - J_N(\boldsymbol{V}_t) \leq -\boldsymbol{v}_t^T\boldsymbol{M}\boldsymbol{v}_t \tag{6.22}$$

holds for all $t \in \mathbb{N}$.

Proof. As shown in Lemma 6.3.4, the shifted sequence $\boldsymbol{V}_{t+1}^{\text{shift}} = \left[\boldsymbol{v}_{t+1|t}^{\text{T}}, \ldots, \boldsymbol{v}_{t+N-1|t}^{\text{T}}, \boldsymbol{0}\right]^{\text{T}}$ generated from the solution of the previous time step $\boldsymbol{V}_t$ is feasible. Evaluating the cost for $\boldsymbol{V}_{t+1}^{\text{shift}}$ leads to $J_N\left(\boldsymbol{V}_{t+1}^{\text{shift}}\right) - J_N\left(\boldsymbol{V}_t\right) = -\boldsymbol{v}_t^{\text{T}}\boldsymbol{M}\boldsymbol{v}_t$ (cf. proof of Lemma 2.3.3). As only feasible solutions with $J_N\left(\boldsymbol{V}_{t+1}\right) \leq J_N\left(\boldsymbol{V}_{t+1}^{\text{shift}}\right)$ are accepted by the algorithm (in step 13-20), it follows that (6.22) holds. This completes the proof. □

Next the set

$$\mathcal{X}_K = \left\{\boldsymbol{x} \in \mathbb{R}^n | \boldsymbol{F}^N\boldsymbol{x} \in \mathcal{X}_{\text{Glo}}^{\text{f}}, \boldsymbol{F}^r\boldsymbol{x} \in \bar{\mathcal{X}}^r, \boldsymbol{K}\boldsymbol{F}^r\boldsymbol{x} \in \bar{\mathcal{U}}^r,\ \forall r \in \mathbb{Z}_{0:N-1}\right\} \tag{6.23}$$

is defined. Note that if $\boldsymbol{x} \in \mathcal{X}_K$, $\boldsymbol{V} = \boldsymbol{0}$ is a feasible solution to the (6.8) with $\mathcal{C}_i = \mathcal{N}_i$ for all $i \in \mathbb{Z}_{1:M}$. Next the algorithm is discussed for the case that $\boldsymbol{x} \in \mathcal{X}_{\text{Glo}}^{\text{f}}$.

Lemma 6.3.6. *Consider optimization problem* (6.8) *with* $\mathcal{C}_{i,0} = \mathcal{N}_i$ *and assume* $\boldsymbol{x}_0 \in \mathcal{X}_{\text{Glo}}^{\text{f}}$*, then the optimal solution is* $\boldsymbol{V}_0 = \boldsymbol{0}$*. Moreover, consider system* (6.12) *and assume* $\boldsymbol{x}_0 \in \mathcal{X}_{\text{Glo}}^{\text{f}}$*, then* $\boldsymbol{x}_t \in \mathcal{X}_K\ \forall t \in \mathbb{N}$.

Proof. The proof is only sketched here due to the similarity to the proof of Lemma 4.3.3. It is first shown that $\mathcal{X}_{\text{Glo}}^{\text{f}} \subseteq \mathcal{X}_K$. From Assumption 6.3.2 it follows that $\mathcal{X}_{\text{Glo}}^{\text{f}}$ is positive invariant for the dynamics $\boldsymbol{x}_{t+1} = \boldsymbol{F}\boldsymbol{x}_t\ \forall t \in \mathbb{N}$. Together with the design of the sets (6.16) and (6.17) it follows that $\mathcal{X}_{\text{Glo}}^{\text{f}} \subseteq \mathcal{X}_K$.

As $\boldsymbol{x}_0 \in \mathcal{X}_{\text{Glo}}^{\text{f}}$ it follows that $\boldsymbol{x}_0 \in \mathcal{X}_K$ and thus $\boldsymbol{V}_0 = \boldsymbol{0}$ is a feasible solution. As J_N has its minimum at $\boldsymbol{V}_0 = \boldsymbol{0}$, it is also the optimal solution. Moreover, as Lemma 6.3.5 holds, it follows that $\boldsymbol{V}_t = \boldsymbol{0}$ are feasible solutions for all $t \in \mathbb{N}$ and thus $\mathcal{X}_K$ is positive invariant for system (6.12). This completes the proof. □

The result concerning recursive feasibility can now be stated.

Theorem 6.3.7. *Consider system* (6.12) *and assume* $\boldsymbol{x}_0 \in \mathcal{X}_N$ *as well as initially feasible solutions* $\boldsymbol{V}_{i,0}$ *for all* $i \in \mathbb{Z}_{1:M}$ *are given, then* $\boldsymbol{x}_t \in \mathcal{X}_N$ *for all* $t \in \mathbb{N}$.

Proof. Let $t_{\text{x}} \in \mathbb{N}$ be the time when $\boldsymbol{x}_t$ enters the set $\mathcal{X}_{\text{Glo}}^{\text{f}}$. $\boldsymbol{x}_t \in \mathcal{X}_N$ for $t < t_{\text{x}}$ follows from Lemma 6.3.4. In step 4-6 of Algorithm 3, $\boldsymbol{V}_t$ is set to zero when $\boldsymbol{x}_t$ enters the set $\mathcal{X}_{\text{Glo}}^{\text{f}}$. As $\mathcal{X}_{\text{Glo}}^{\text{f}} \subseteq \mathcal{X}_K$ and $\mathcal{X}_K$ is an RPI set for (6.12), it follows $\boldsymbol{x}_t \in \mathcal{X}_K$ as shown in Lemma 6.3.6. Since $\mathcal{X}_K \subseteq \mathcal{X}_N$, $\boldsymbol{x}_t \in \mathcal{X}_N$ for all $t \in \mathbb{N}$. This completes the proof. □

Note that the theorem assumes initially feasible sequences $\boldsymbol{V}_{i,0}$ for all $i \in \mathbb{Z}_{1:M}$ exist. They can for instance be obtained by setting $J_N\left(\boldsymbol{V}_0^{\text{shift}}\right) := \infty$ and applying Algorithm 3. Any feasible solution found by the algorithm is accepted.

Remark 6.3.8. *For the case that $\mathcal{W}$, $\mathcal{U}$ and $\mathcal{X}$ are polyhedra, the constraint tightened sets $\bar{\mathcal{U}}_i^l$ and $\bar{\mathcal{X}}_i^l$ are not more complex than the constraints $\mathcal{U}_i$ and $\mathcal{X}_i$. This also holds for the ellipsoidal terminal constraint sets $\mathcal{T}_i$ and $\mathcal{X}_i^f$. Thus, the computational complexity of* (6.8) *is not higher that the one of* (6.6).

6.3.2 Stability Analysis

The results concerning stability are stated in the next theorem.

Theorem 6.3.9. *Consider the closed-loop system* (6.12) *and let $\mathcal{Z}$ be the minimal RPI set for the system $\boldsymbol{x}_{t+1} = \boldsymbol{F}\boldsymbol{x}_t + \boldsymbol{w}_t$. For all $\boldsymbol{x}_0 \in \mathcal{X}_N$ and given initially feasible solutions $\boldsymbol{V}_{i,0}$ for all $i \in \mathbb{Z}_{1:M}$, it follows that $\lim_{t\to\infty} \|\boldsymbol{x}_t\|_{\mathcal{Z}} = 0$, i.e., $\boldsymbol{x}_t$ converges to the set $\mathcal{Z}$ for $t \to \infty$. If, furthermore, there exists an $\epsilon \in \mathbb{R}_{>0}$ such that $\mathcal{Z} \oplus \mathcal{B}_\epsilon \subseteq \mathcal{X}^{\mathrm{f}}_{\mathrm{Glo}}$, then the set $\mathcal{Z}$ is GES with ROA $\mathcal{X}_N$.*

Proof. GES of the set $\mathcal{Z}$ follows from Theorem 3.4.2. Thereby, attractivity of the set $\mathcal{Z} \times \{\boldsymbol{0}\}$ directly follows from Lemma 3.4.1. Note that $J_N(\boldsymbol{V}_{t+1}) - J_N(\boldsymbol{V}_t) \leq -\boldsymbol{v}_t^{\mathrm{T}}\boldsymbol{M}\boldsymbol{v}_t$ holds due to Lemma 6.3.5 and $\boldsymbol{F}^{\mathrm{T}}\boldsymbol{P}\boldsymbol{F} - \boldsymbol{P} \preceq -\boldsymbol{Q}_K$ holds due to Assumption 6.1.1.

Equation (3.9) holds by $\mathcal{Z} \oplus \mathcal{B}_\epsilon(\boldsymbol{0}) \subseteq \mathcal{X}^{\mathrm{f}}_{\mathrm{Glo}}$. This completes the proof. □

6.4 Illustrative Example

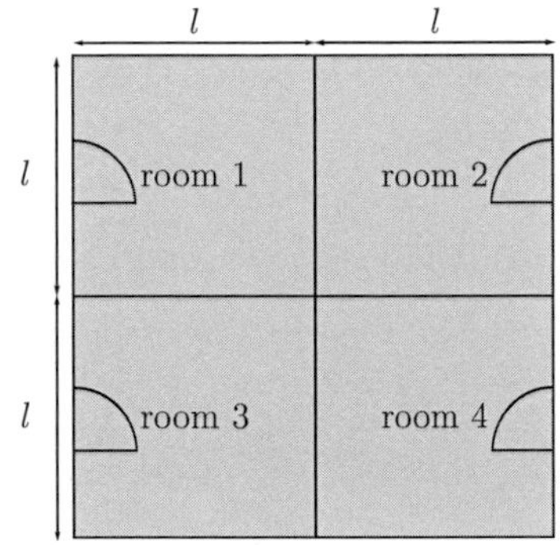

Figure 6.2: Four room building

The event-triggered cooperation approach is evaluated for the control of the temperatures in the four room building shown in Figure 6.2. The set-up is taken from [BFS14]. The control objective is to steer the temperature T_i in each room to a desired value $\bar{T}$. Each room corresponds to a subsystem and is equipped with a radiator which supplies the heat q_i. The outside temperature is $T_e = 0\,°\mathrm{C}$. The system data can be found in Table 6.1. The equilibrium point is $\bar{T}_1 = \bar{T}_2 = \bar{T}_3 = \bar{T}_4 = \bar{T} = 20\,°\mathrm{C}$ and $\bar{q}_1 = \bar{q}_2 = \bar{q}_3 = \bar{q}_4 = \bar{q} = \bar{T} s_{\mathrm{e}} k_{\mathrm{e}}$. Defining the state and input as the differences of the equilibrium values, i.e.,

$$\boldsymbol{x} = \begin{pmatrix} \delta T_1 & \delta T_2 & \delta T_3 & \delta T_4 \end{pmatrix}^{\mathrm{T}} \text{ and } \boldsymbol{u} = \begin{pmatrix} \delta q_1 & \delta q_2 & \delta q_3 & \delta q_4 \end{pmatrix}^{\mathrm{T}}, \tag{6.24}$$

respectively, where $\delta T_i = T_i - \bar{T}$ and $\delta q_i = \frac{(q_i - \bar{q})}{cV\rho}$, the dynamic model is given by

$$\dot{\boldsymbol{x}}(t) = \boldsymbol{A}_{\mathrm{c}} \boldsymbol{x}(t) + \boldsymbol{B}_{\mathrm{c}} \boldsymbol{u}(t) + \boldsymbol{w}(t) \tag{6.25}$$

$$\boldsymbol{A}_{\mathrm{c}} = \begin{pmatrix} -\sigma & \sigma_2 & \sigma_1 & 0 \\ \sigma_2 & -\sigma & 0 & \sigma_1 \\ \sigma_1 & 0 & -\sigma & \sigma_2 \\ 0 & \sigma_1 & \sigma_2 & -\sigma \end{pmatrix} \quad \boldsymbol{B}_{\mathrm{c}} = \begin{pmatrix} 1 & 0 & 0 & 0 \\ 0 & 1 & 0 & 0 \\ 0 & 0 & 1 & 0 \\ 0 & 0 & 0 & 1 \end{pmatrix}$$

where $\sigma_1 = \frac{s_{\mathrm{r}} k_1}{c\rho V}$, $\sigma_2 = \frac{s_{\mathrm{r}} k_2}{c\rho V}$, $\sigma_{\mathrm{e}} = \frac{s_{\mathrm{e}} k_{\mathrm{e}}}{c\rho V}$ and $\sigma = \sigma_1 + \sigma_2 + \sigma_{\mathrm{e}}$. The constraints on the state $\boldsymbol{x}_i$ and the inputs $\boldsymbol{u}_i$ are $\boldsymbol{x}_i \in \mathcal{X}_i = \{\boldsymbol{x}_i \in \mathbb{R} | -5 \leq \boldsymbol{x}_i \leq 5\}$ and $\boldsymbol{u}_i \in \mathcal{U}_i = \{\boldsymbol{u}_i \in \mathbb{R} | -0.038 \leq \boldsymbol{u}_i \leq 0.03\}$ for all $i \in \mathbb{Z}_{1:M}$ with $M = 4$. The disturbance sets are

Table 6.1: System parameter of four room building

description	name	value
air density	ρ	$1.2\,\mathrm{kg/m^3}$
air heat capacity	c	$1005\,\mathrm{J/kgK}$
length room	l	$4\,\mathrm{m}$
volume room	V	$48\,\mathrm{m}^3$
wall surface between rooms	s_r	$12/\mathrm{m}^2$
external wall surface	s_e	$24/\mathrm{m}^2$
heat transfer coeff. 1-3, 2-4	k_1	$1\,\mathrm{W/m^2K}$
heat transfer coeff. 1-2, 3-4	k_2	$2.5\,\mathrm{W/m^2K}$
heat transfer coeff. ext.	k_e	$0.5\,\mathrm{W/m^2K}$
sampling interval	h	$10\,\mathrm{s}$

$\mathcal{W}_i = \{w_i \in \mathbb{R} | |w_i| \leq 0.05\}$ for all $i \in \mathbb{Z}_{1:M}$. The values for $\boldsymbol{w}_{i,t}$ are chosen randomly in $\mathcal{W}_i$ using an uniform distribution. For simulation the discrete-time model of (6.25) computed using Euler-forward with a fixed sampling interval h is utilized. The matrices $\boldsymbol{P}_i$, $\boldsymbol{K}_{\mathcal{N}_i}$, $\boldsymbol{T}_{\mathcal{N}_i}$ and β are designed using LMIs as described in [CJMZ16] with $\boldsymbol{Q}_K = \boldsymbol{Q} + \boldsymbol{K}^\mathrm{T}\boldsymbol{R}\boldsymbol{K}$ where $\boldsymbol{Q} = \mathrm{diag}_{i\in\mathbb{Z}_{1:M}}(\boldsymbol{Q}_i)$, $\boldsymbol{Q}_i = 1$ and $\boldsymbol{R} = \mathrm{diag}_{i\in\mathbb{Z}_{1:M}}(\boldsymbol{R}_i)$, $\boldsymbol{R}_i = 1$ for all $i \in \mathbb{Z}_{1:M}$. The weighting matrices are chosen to $\boldsymbol{M}_i = \boldsymbol{R}_i + \boldsymbol{B}_{ii}^\mathrm{T}\boldsymbol{P}_i\boldsymbol{B}_{ii}$. The constraint and terminal sets are designed as described in Section 6.3.1. To solve the distributed optimization problem the ADMM approach of [BT97, Chapter 2] is used. For the value of $\rho = 0.05$ (ADMM tuning parameter) the algorithm converged for all considered cases. Moreover, the event-triggering sets $\mathcal{E}_i^\mathrm{x} = \{x_i \in \mathbb{R} | |x_i| \leq 0.03\}$ for all $i \in \mathbb{Z}_{1:M}$ are selected. The triggering and stopping conditions are checked every 5 steps ($\Delta k = 5$).

Figure 6.3 shows the simulation results for two different initial values. As expected, the state approaches the origin and remains close to it as well as the constraints are not violated. In the lower plot, one can observe that when the state is far away from the origin there is more cooperation than when the state is close to the origin. One can further see that the controllers adapt the cooperation structure in the system since mostly the controllers of the subsystems whose states are far away from the origin cooperate with their neighbors. In total the controllers cooperate 30 times. The algorithm finds a feasible and cost-improving solution in maximum 10 iterations.

Next, the event-triggered cooperation approach (ETC approach) is compared to a fully cooperating approach (FC approach) where Algorithm 3 is applied with $\mathcal{E}_i^\mathrm{x} := \emptyset$ and $\mathcal{C}_{i,t}^0 := \mathcal{N}_i$ for all $t \in \mathbb{N}$. The simulation is carried out for 100 initial states $\boldsymbol{x}_0$ randomly chosen in the constraint set and the closed-loop is simulated for $T_\mathrm{sim} = 25$ steps for each initial state. In average the controllers cooperate 31.95 times for the ETC approach and $4{\cdot}T_{\mathrm{sim}} = 100$ times for the FC approach. The average number of iterations when cooperating can be seen as a measure for the communication effort. For the FC approach it is 1033 and for the ETC scheme 349.9. Evidently, the number of messages sent by the ETC approach is fewer. The average of the closed-loop costs $J = \sum_{t=0}^{T_{\mathrm{sim}}} \boldsymbol{x}_t^{\mathrm{T}} \boldsymbol{Q} \boldsymbol{x}_t + \boldsymbol{u}_t^{\mathrm{T}} \boldsymbol{R} \boldsymbol{u}_t$ for both approaches are $J_{\mathrm{FC}} = 127.633$ and $J_{\mathrm{ETC}} = 127.634$. This shows that the ETC approach is performing almost equivalent to the FC approach for the considered set-up.

6.5 Summary

In this chapter a decomposition-based DMPC algorithm for constrained linear dynamically coupled subsystems with additive disturbance is introduced. The algorithm employs an event-triggered cooperation idea to decide when the controllers of the subsystems have to cooperate with the neighboring controllers. As a result the cooperation structure is adapted to the current status of the system and the controllers do not need to cooperate in every time instant with all the neighbors. This leads to a reduction of the overall communication effort in the system. The events for a common cooperation are triggered when the predicted states deviate too much from previously transmitted state values, i.e., when an absolute threshold is violated.

A specially designed structured time-varying terminal set approach is introduced to account for the disturbances as well as the errors caused by the event-triggered and thus time-varying cooperation structure. At the same time, the structural properties of the underlying DMPC problem are preserved. Sufficient conditions are provided and it is discussed how the terminal set can be designed and updated online.

A stopping criterion which excepts suboptimal feasible solutions is included in the algorithm. With the help of the Lyapunov function derived in Section 3.4, it can be shown that the state converges to the minimum RPI set of the system and sufficient conditions for establishing stability of the set are given.

The simulation results show that through the time-varying cooperation, the overall communication effort is reduced while the performance degradation caused by the event-triggered cooperation is relatively small compared to a fully cooperating algorithm.

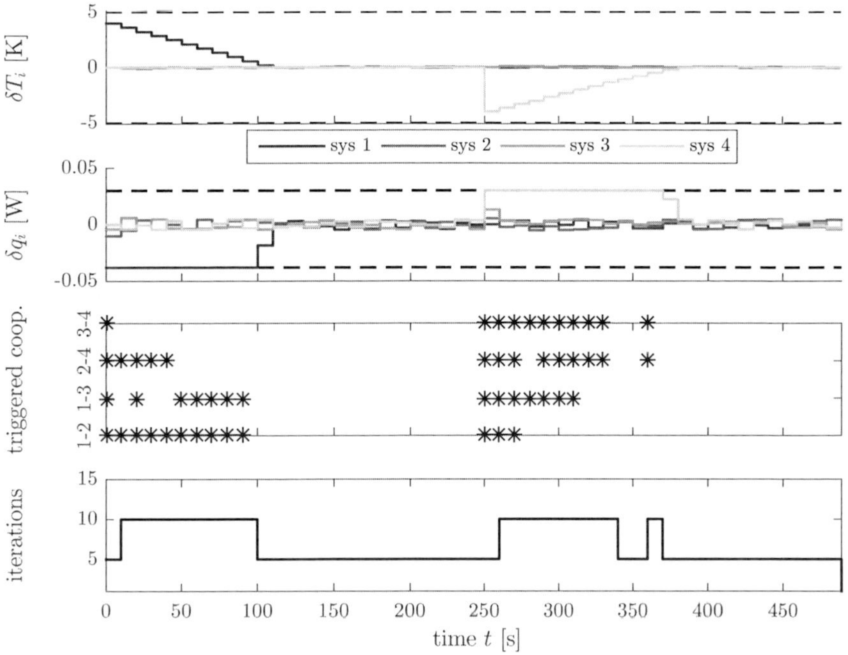

Figure 6.3: The state and input trajectories obtained from simulation with the discrete-time model of the four room building (6.25) are shown in the first and second plot. The trajectories start from $\delta T_{1,0} = 4\,\mathrm{K}$, $\delta T_{i,0} = 0\,\mathrm{K}$ for $i \in \{2, 3, 4\}$. At time $t = 25$ the temperature in room 4 decreases abruptly to $\delta T_{4,25} = -4\,\mathrm{K}$. The constraints are depicted as dashed lines. The asterisks in the third plot indicate if a cooperation has been triggered between two neighboring controllers at the sampling instants. In the lower plot the number of iterations are shown.

Each subsystem is controlled by a local controller which can measure the local state $\boldsymbol{x}_i$ and can communicate with the other subsystems over a communication network. Assumptions 3.3.1 and 3.3.2 hold, i.e., the components in the subsystems as well as all the subsystems are synchronized and the communication network is ideal.

The cost of every subsystem $\mathscr{S}_i$ is

$$V_{N,i,t}\left(\boldsymbol{X}_{i,t}, \boldsymbol{U}_{i,t}\right) = \sum_{r=t}^{t+N-1} l_{i,r}\left(\boldsymbol{x}_{i,r}, \boldsymbol{u}_{i,r}\right) \tag{7.4}$$

where $l_{i,t} : \mathbb{R}^{n_i} \times \mathbb{R}^{m_i} \to \mathbb{R}$ are local stage costs which are continuous and periodically time-varying, i.e., $l_{i,t}(\cdot,\cdot) = l_{i,t+P}(\cdot,\cdot)$ for all $t \in \mathbb{N}$ and $i \in \mathbb{Z}_{1:M}$. They do not need to be positive definite for some reference value and ideally reflect an economical objective.

The goal of this section is to design a DMPC scheme for the dynamics (7.1) where the controllers of the subsystems minimize the possibly economical objective function cooperatively while satisfying the constraints (7.2) and (7.3). Therefore, the controllers use an iterative algorithm that is tailored to the coupling structure and the constraints. To keep the communication effort low, they exchange predicted state and input sequences in an event-triggered fashion.

7.1.2 Distributed Control Scheme

As outlined in Section 2.3, the optimal periodic behavior of the global system is described by the periodic sequences $\boldsymbol{X}^{\mathrm{P}}$ and $\boldsymbol{U}^{\mathrm{P}}$ which can be determined by solving the problem (2.24) offline. Thereby, the global stage cost is $l_r(\boldsymbol{x}, \boldsymbol{u}) = \sum_{i=1}^{M} l_{i,r}(\boldsymbol{x}_i, \boldsymbol{u}_i)$. For more information on how to obtain the required global model it is referred to Subsection 3.2.2.

Before the algorithm is discussed, the following notation is introduced. $\mathcal{N}_i^{\mathrm{U}}$ is the set of inputs over which the controller of subsystem $\mathscr{S}_i$ optimizes. Note that through the dynamical coupling the states in $\mathcal{N}_i^{\mathrm{X}} = \{j \in \mathbb{Z}_{1:M} | i \in \mathcal{N}_j\} \cup \{i\}$ are directly affected. Additionally, the set $\mathcal{N}_i^{Z} := \left\{s \in \mathbb{Z}_{1:n_c} | s \in \mathcal{N}_r^{\mathrm{CX}} \cup \mathcal{N}_r^{\mathrm{CU}}, \forall r \in \mathcal{N}_i^{\mathrm{U}} \cup \mathcal{N}_i^{\mathrm{X}}\right\}$ is defined. It is set of constraints that include an input over which subsystem $\mathscr{S}_i$ optimizes or a state which is directly affected by it.

At the core of the DMPC algorithm, the following optimization problem is solved in every subsystem $\mathscr{S}_i$ with $i \in \mathbb{Z}_{1:M}$:

$$\underset{\{\boldsymbol{U}_{j,t}\}, j \in \mathcal{N}_i^{\mathrm{U}}}{\text{minimize}} \quad \sum_{j \in \mathcal{N}_i^{\mathrm{U}} \cup \mathcal{N}_i^{\mathrm{X}}} V_{N,j,t}\left(\boldsymbol{X}_j, \boldsymbol{U}_j\right) \tag{7.5a}$$

$$\text{subject to} \quad \text{for all } j \in \mathcal{N}_i^{\mathrm{X}}:$$

$$\boldsymbol{x}_{j,t|t} = \boldsymbol{x}_{j,t} \tag{7.5b}$$

$$\boldsymbol{x}_{j,t+N|t} = \boldsymbol{x}^{\mathrm{P}}_{j,[t+N]_P} \tag{7.5c}$$

for all $r \in \mathbb{Z}_{t:t+N-1}$:

$$\boldsymbol{x}_{j,r+1|t} = \boldsymbol{A}_{jj,r}\boldsymbol{x}_{j,r|t} + \boldsymbol{B}_{jj,r}\boldsymbol{u}_{j,r|t} + \sum_{s\in\mathcal{N}_j} \boldsymbol{B}_{js}\boldsymbol{u}_{s,r|t} + \boldsymbol{d}_{j,r}, \tag{7.5d}$$

$$\boldsymbol{x}_{j,r|t} \in \mathcal{X}_{j,t} \tag{7.5e}$$

for all $j \in \mathcal{N}_i^{\mathrm{U}}$:

$$\boldsymbol{u}_{j,r|t} \in \mathcal{U}_{j,t} \tag{7.5f}$$

for all $j \in \mathcal{N}_i^{Z}$:

$$\left(\mathrm{col}_{s\in\mathcal{N}_j^{\mathrm{CX}}}\left(\boldsymbol{x}_{s,t}\right), \mathrm{col}_{s\in\mathcal{N}_j^{\mathrm{CU}}}\left(\boldsymbol{u}_{s,t}\right)\right) \in \mathcal{Z}^{\mathrm{C}}_{j,t}, \tag{7.5g}$$

for all $j \in (\bigcup_{s\in\mathcal{N}_i^{\mathrm{Z}}} \mathcal{N}_s^{\mathrm{CU}} \cup \bigcup_{s\in\mathcal{N}_i^{\mathrm{X}}} \mathcal{N}_s) \setminus \mathcal{N}_i^{\mathrm{U}}$:

$$\boldsymbol{u}_{j,r|t} = \breve{\boldsymbol{u}}^{k}_{j,r} \tag{7.5h}$$

for all $j \in (\bigcup_{s\in\mathcal{N}_i^{\mathrm{Z}}} \mathcal{N}_s^{\mathrm{XU}}) \setminus \mathcal{N}_i^{\mathrm{X}}$:

$$\boldsymbol{x}_{j,r|t} = \breve{\boldsymbol{x}}^{k}_{j,r}. \tag{7.5i}$$

Starting from the measurements $\boldsymbol{x}_{j,t}$ in (7.5b), each subsystem predicts the future states of systems that are directly influenced using the system model (7.1) as constraint (7.5d). The constraint (7.5c) ensures that the terminal states lie on the periodic optimal sequence. Moreover, the subsystem has to consider state, input and mixed constraints in (7.5e-g). For inputs from the subsystems that are involved in one of the constraints (7.5g) or couplings (7.5d) but which is not optimized over, input values received from other subsystems are taken and included in constraint (7.5h) into the optimization problem. The states from subsystems involved in the coupling constraint (7.5g) and which are not directly affected by the local input $\boldsymbol{u}_i$ are included into the optimization in (7.5i). The $\breve{(\cdot)}$ refers to inputs and states that it is not optimized over in subsystem $\mathscr{S}_i$ but the values are required for prediction in (7.5d) or constraint satisfaction in (7.5g). They are received from other subsystems via communication.

The cost (7.5a) includes the local cost as well as the cost of subsystems $\mathscr{S}_j$ which is directly influenced by the inputs over which it is optimized, i.e., $j \in \mathcal{N}_i^{\mathrm{X}} \cup \mathcal{N}_i^{\mathrm{U}}$. Due to this the DMPC scheme works cooperatively. $\boldsymbol{X}_i$ and $\boldsymbol{U}_i$ are the local predicted state and input sequences.

Although the optimization problem and the algorithm are presented for a quite general set-up, it is assumed that there is some special structure in the constraints or coupling so that the dimensions of the optimization problems do not become too large. This becomes more obvious in the considered applications where cascaded or decoupled dynamics as well as cascaded or fully coupled constraints are studied.

Next, the steps of the iterative DMPC Algorithm 4 are discussed. $k \in \mathbb{Z}_{1:k_{\max}}$ is the iteration counter and $k_{\max}$ is the maximum number of iterations. At the beginning each controller solves the DMPC problem (7.5) locally having state and input sequences $\check{\boldsymbol{X}}_j$ and $\check{\boldsymbol{U}}_j$ for the required subsystems from the last iteration or initialization. The optimizers have a superscript $[i]$ so that one can distinguish between the solutions of the subsystems. Then, the difference ΔV_i of the optimized cost to the cost of the previous iteration is computed. If the difference is smaller than some predefined threshold $\eta_i \in \mathbb{R}_{\leq 0}$, an event is generated. This implies that the binary trigger variable $\gamma_i \in \mathbb{Z}_{0:1}$ is set to 1, otherwise it is 0. Notably, this type of threshold utilized in step 3 has also been applied in [GS13a, GS16]. From the subsystems which triggered an event, the subsystem with the biggest cost decrease is determined in step 4 and called subsystem $\mathscr{S}_{i^*}$. For this either global communication is required or it is evaluated iteratively by N2N communication. In step 5 $\boldsymbol{U}_j^{[i^*]}$ is sent to the subsystems where a predicted state is directly affected or to subsystems which require $\boldsymbol{U}_j^{[i^*]}$ in constraint (7.5g), i.e., to the subsystems $\mathscr{S}_j$ with $j \in \{s \in \mathbb{Z}_{1:M} | i^* \in \mathcal{N}_s^{\mathrm{X}} \wedge i^* \in \mathcal{N}_r^{\mathrm{CU}}, r \in \mathcal{N}_s^{\mathrm{Z}}\}$. Analogously, the state sequence $\boldsymbol{X}_{i^*}$ is sent subsystems $\mathscr{S}_j$ with $j \in \{s \in \mathbb{Z}_{1:M} | i^* \in \mathcal{N}_s^{\mathrm{X}} \wedge i^* \in \mathcal{N}_r^{\mathrm{CX}}, r \in \mathcal{N}_s^{\mathrm{Z}}\}$, respectively. Note that the communication of input and state sequences is only required when an event is triggered. Thereby, the communication effort is kept low. The solutions are then included in the corresponding systems and the solutions of the remaining systems are discarded in step 6. The procedure is repeated until the maximum number of iterations is reached or no more event is triggered, i.e., $\mathcal{I} = \emptyset$ (step 7). After that the initial state and input sequences for the next time step are generated by shifting the current state and input sequence and extending it with the corresponding state or input of the optimal sequences in step 8. This step can be executed without communication if the optimal sequence is known in all subsystems. In the last step the first input is applied to the system.

Note that in each iteration the algorithm has feasible iterates and the costs are non-increasing over the iterations. A feasible solution for the next time step is ensured by the shifting and extension using the optimal sequence. For time $t = 0$ initially feasible input sequences are required. The procedure for determining them is application dependent and is described in the corresponding sections.

Next, the theoretical properties are discussed. The set of states for which a feasible input exists at time t is called $\mathcal{X}_{N,t}$.

Theorem 7.1.1. *Consider system* (7.1) *under the control law obtained from Algorithm 4 and assume that* $\boldsymbol{x}_0 \in \mathcal{X}_{N,0}$, *then* $\boldsymbol{x}_t \in \mathcal{X}_{N,t}$ *for all* $t \in \mathbb{N}$.

Proof. First, recursive feasibility is considered. It is shown that starting from feasible

Algorithm 4 DMPC algorithm for LPTV subsystems

1: **Measure** $\boldsymbol{x}_{i,t}$ and send it to subsystems $\mathscr{S}_j$ with $j \in \mathcal{N}_i^{\mathrm{X}}$

2: **for** $k = 1, \dots, k_{\max}$

3: For each Subsystem $\mathscr{S}_i$ in parallel

Solve (7.5) to obtain $\left\{\boldsymbol{U}_{j,t}^{[i]}\right\}_{j \in \mathcal{N}_i^{\mathrm{U}}} = \arg\min_{\{\boldsymbol{U}_{j,t}\}, j \in \mathcal{N}_i^{\mathrm{U}}} \sum_{j \in \mathcal{N}_i^{\mathrm{X}} \cup \mathcal{N}_i^{\mathrm{U}}} V_{N,j,t}\left(\boldsymbol{X}_j, \boldsymbol{U}_j\right)$

Construct $\boldsymbol{X}_j^{[i]}$, $j \in \mathcal{N}_i^{\mathrm{X}}$

Compute $\Delta V_i = \sum_{j \in \mathcal{N}_i^{\mathrm{X}} \cup \mathcal{N}_i^{\mathrm{U}}} V_{N,j,t}\left(\boldsymbol{X}_j^{[i]}, \boldsymbol{U}_j^{[i]}\right) - V_{N,j,t}\left(\check{\boldsymbol{X}}_{j,t}^{k}, \check{\boldsymbol{U}}_{j,t}^{k}\right)$

Evaluate event-trigger: **if** $\Delta V_i < \eta_i$ **then** $\gamma_i = 1$ **else** $\gamma_i = 0$

4: **Find**: $i^* = \arg\min_{i \in \mathcal{I}} \Delta V_i$ with $\mathcal{I} = \{i \in \mathbb{Z}_{1:M} | \gamma_i = 1\}$

5: **Send** $\boldsymbol{X}_j^{[i^*]}$, $j \in \mathcal{N}_{i^*}^{\mathrm{X}}$ and $\boldsymbol{U}_j^{[i^*]}$, $j \in \mathcal{N}_{i^*}^{\mathrm{U}}$ from subsystem $\mathscr{S}_{i^*}$

6: **for** $i = 1, \dots, M$

if $i \in \mathcal{N}_{i^*}^{\mathrm{U}}$ **then** $\check{\boldsymbol{U}}_{i,t}^{k+1} = \boldsymbol{U}_i^{[i^*]}$ **else** set $\check{\boldsymbol{U}}_{i,t}^{k+1} = \check{\boldsymbol{U}}_{i,t}^{k}$,

if $i \in \mathcal{N}_{i^*}^{\mathrm{X}}$ **then** $\check{\boldsymbol{X}}_{i,t}^{k+1} = \boldsymbol{X}_i^{[i^*]}$ **else** set $\check{\boldsymbol{X}}_{i,t}^{k+1} = \check{\boldsymbol{X}}_{i,t}^{k}$

endfor

7: **if** $\mathcal{I} = \emptyset$ **then** go to step 8

endfor

8: **for** $i = 1, \dots, M$

Set $\check{\boldsymbol{U}}_{i,t+1}^{1} = \left[\check{\boldsymbol{u}}_{i,t+1|t}^{k+1,\mathrm{T}}, \dots, \check{\boldsymbol{u}}_{i,t+N-1|t}^{k+1,\mathrm{T}}, \boldsymbol{u}_{i,[t+N]_P}^{\mathrm{P,T}}\right]^{\mathrm{T}}$,

Set $\check{\boldsymbol{X}}_{i,t+1}^{1} = \left[\check{\boldsymbol{x}}_{i,t+1|t}^{k+1,\mathrm{T}}, \dots, \check{\boldsymbol{x}}_{i,t+N|t}^{k+1,\mathrm{T}}, \boldsymbol{x}_{i,[t+N]_P+1}^{\mathrm{P,T}}\right]^{\mathrm{T}}$

endfor

9: **Set** $\boldsymbol{u}_t = \mathrm{col}_{i \in \mathbb{Z}_{1:M}}\left(\check{\boldsymbol{u}}_{i,t}^{k+1}\right)$

sequences $\check{\boldsymbol{U}}_{i,t}^{1}$ and $\check{\boldsymbol{X}}_{i,t}^{1}$ $\forall i \in \mathbb{Z}_{1:M}$ at time t, the sequences $\check{\boldsymbol{U}}_{i,t}^{k}$ $\forall i \in \mathbb{Z}_{1:M}$ remain feasible over the iterations. Then, it is shown how the sequences $\check{\boldsymbol{U}}_{i,t+1}^{1}$ and $\check{\boldsymbol{X}}_{i,t+1}^{1}$ can be constructed for the time $t+1$.

Assume that $\boldsymbol{x}_t \in \mathcal{X}_{N,t}$ and $\check{\boldsymbol{U}}_{i,t}^{1}$ $\forall i \in \mathbb{Z}_{1:M}$ as well as $\check{\boldsymbol{X}}_{i,t}^{1}$ $\forall i \in \mathbb{Z}_{1:M}$ are feasible solutions to the optimization problems (7.5). This implies that there exist at least one set of solutions such that the problems are feasible and thus the sequences $\boldsymbol{U}_{j,t}^{[i]}$ with $j \in \mathcal{N}_i^{\mathrm{U}}$, $\check{\boldsymbol{U}}_{j,t}^{1}$ with $j \in \mathbb{Z}_{1:M} \setminus \mathcal{N}_i^{\mathrm{U}}$, $\boldsymbol{X}_{j,t}^{[i]}$ with $j \in \mathcal{N}_i^{\mathrm{X}}$ and $\check{\boldsymbol{X}}_{j,t}^{1}$ with $j \in \mathbb{Z}_{1:M} \setminus \mathcal{N}_i^{\mathrm{X}}$ are feasible for all $i \in \mathbb{Z}_{1:M}$. Through the assigning step 6, $\check{\boldsymbol{U}}_{i,t}^{2}$ and $\check{\boldsymbol{X}}_{i,t}^{2}$ $\forall i \in \mathbb{Z}_{1:M}$ are feasible as only one solution is accepted (sequential update). By induction it follows that $\check{\boldsymbol{U}}_{i,t}^{k}$ and $\check{\boldsymbol{X}}_{i,t}^{k}$ $\forall i \in \mathbb{Z}_{1:M}$ are feasible for $k \in \mathbb{Z}_{1:k_{\max}}$.

Next it is shown that the sequences $\check{\boldsymbol{U}}^1_{i,t+1}$, $\check{\boldsymbol{X}}^1_{i,t+1}$ generated in step 8 are feasible for the optimization problem (7.5). Since $\check{\boldsymbol{U}}^k_{i,t}$ and $\check{\boldsymbol{X}}^k_{i,t}$ $\forall i \in \mathbb{Z}_{1:M}$ are feasible as well as $\boldsymbol{x}_{i,t+1} = \boldsymbol{x}^{k+1}_{i,t+1|t}$, $\boldsymbol{u}^{k+1}_{i,t+r|t}$ and $\boldsymbol{x}^{k+1}_{i,t+r|t}$ satisfy the constraints for $r \in \mathbb{Z}_{1:N-1}$. As $\boldsymbol{x}^{k+1}_{i,t+N|t} = \boldsymbol{x}^{\mathrm{P}}_{i,[t+N]_P}$, it follows from the optimal sequence that $\boldsymbol{x}^{\mathrm{P}}_{i,[t+N]_P}$,$\boldsymbol{u}^{\mathrm{P}}_{i,[t+N]_P}$ and $\boldsymbol{x}^{\mathrm{P}}_{i,[t+N]_P+1}$ satisfy the constraints. Thus, $\check{\boldsymbol{U}}^1_{i,t+1}$ and $\check{\boldsymbol{X}}^1_{i,t+1}$ are feasible solutions for (7.5) and $\boldsymbol{x}_{t+1} \in \mathcal{X}_{N,t+1}$.

□

Remark 7.1.2. *In the proof of Theorem 7.1.1 it is assumed that the measurements at time t match the predicted value for* $t+1$ *at time t, i.e.,* $\boldsymbol{x}_{i,t+1} = \boldsymbol{x}^{k+1}_{i,t+1|t}$. *Due to unpredicted disturbances or modeling errors, this might not hold in a practical application and thus a robust controller must be designed for guaranteeing recursive feasibility. However, the design of a robust MPC controller for the considered set-up is a challenging task as it requires the computation of (robustly) positive invariant sets. Due to the integer nature of the problem this is not straightforward. A "practical" approach to maintain recursive feasibility is discussed in the simulation section.*

Remark 7.1.3. *Note that compared to a centralized controller, a distributed one is usually more robust to failures in the system. If the centralized controller fails, the overall plant cannot be controlled anymore. For the considered DMPC algorithm, however, it might in some cases be possible to control at least some of the subsystems in case of a failure in one of the controllers. In which cases this is possible dependents on the application. Examples for the considered set-ups are discussed in the next sections. For failure situations the optimal periodic sequence changes. They can be computed offline before operation and stored for the online usage. Notably, this advantage is of particular importance when infrastructure systems, such as WDNs or the power grids, are considered where the safety of supply has a high priority.*

7.2 Application to Control of a WDN

WDNs transport drinking water from supply systems to the consumers. The systems are usually large-scale, complex, constrained and highly nonlinear, for instance, due to losses in the pipes and pump characteristics. Moreover, the systems have a non-continuous character as the pumps are often switched on/off since the water demand can vary strongly over the day. Considering several days, the water demand does not change significantly from day to day, i.e., the WDNs have a periodic character. The major requirement for the operational control of a WDN is that the water demand in the system is covered at all times as water is vital for humans. The supply must also be guaranteed for safety reasons, e.g., in the case of fire. Further requirements are low operational costs.

Table 7.1: Overview of literature on DMPC approaches for WDNs

Paper	Losses	Pump characteristic	Fixed-speed pumps	Control architecture	Objective
[OBPB12]	-	-	-	hierarchical: centralized, decentralized	economic, regulation
[LZNS10]	✓	✓	-	distributed	economic
[GOMP17]	-	-	-	distributed	economic

Due to the high dimensionality and the complexity of WDNs as well as the high performance and safety requirements in a constrained environment, DMPC is a well-suited control method. Applications of DMPC to WDNs are investigated, e.g., in [LZNS10, OBPB12, GOMP17].

In this section the DMPC scheme presented in Section 7.1 is tested for a set-up derived from a part of the WDN in the city of Kaiserslautern. The WDN is located in a mountainous region and divided into several pressure zones which are interconnected by fixed-speed pumps. The water is taken from a spring and 20 deep wells in two different supply zones. The WDN has a total length of approx. 454 km and delivers water to approx. 20,000 houses. In total there are 13 reservoirs with a storage capacity of approx. 17,000 m^3. The data stems from a local operator.

First, a nonlinear model derived from the WDN in Kaiserslautern is described. From that, a control-oriented model is developed. The model is tailored to the special topology in the system and considers the pressures, the presence of fixed-speed pumps and losses. As shown in Table 7.1, those significant properties are usually neglected in DMPC applications for WDNs. For the inclusion of the characteristics of the fixed-speed pumps a polytopic approximation is introduced which is sufficient accurate and the resulting controller optimization problem can be handled efficiently from a computational point of view. Finally, the control scheme presented in Section 7.1 is tested for the WDN.

7.2.1 Modeling of WDNs

Modeling of components in a WDN

Next, the general modeling of components in a WDN which are required for further investigations is reviewed. If not stated otherwise, this section follows the presentation in [BU94, chapter 2].

A WDN can be described by a graph consisting of nodes and branches. The branches can be pipes, valves or pump stations and link the nodes with each other. Based on the flow continuity law for all nodes i

$$\sum_{j\in\mathcal{N}_i} q_{ij} = d_{\mathrm{W}i} \tag{7.6}$$

holds where the set $\mathcal{N}_i$ consists of the indexes of the neighbor nodes of node i, i.e., the nodes where there is a branch to. The flows in the branch are q_{ij} and $d_{\mathrm{W}i}$ is the demand in this node. The hydraulic head h_i (which corresponds to the pressure and is also called head in the sequel) in the node i is given by

$$h_i = E_i + p_i, \tag{7.7}$$

i.e., the sum of elevation E_i and the water pressure head p_i.

Losses in a pipe are given in terms of a head drop by the Hazen-Williams formula

$$\Delta h_{\mathrm{p}ij} = R_{ij} q_{ij} |q_{ij}|^{0.852} \tag{7.8}$$

where R_{ij} is the pipe resistance. This empirical relationship can be computed based on the pipe length, diameter and roughness coefficient if available.

The head difference between two nodes connected with a branch consisting of a pump station with n_{ij}^{P} identical variable speed pumps operated at the same pump speed ω_{ij} is given by the characteristic

$$\Delta h_{\mathrm{s}ij} = \begin{cases} \left[\frac{c_{2ij}}{\delta_{ij}}\right] q_{ij}^2 + \left[\frac{c_{1ij}}{\delta_{ij}}\right] \omega_{ij} q_{ij} + c_{0ij}\omega_{ij}^2, & \delta_{ij} \neq 0 \\ h_j - h_i, & \delta_{ij} = 0 \end{cases} \tag{7.9}$$

with $q_{ij} \geq 0$, the number of running pumps δ_{ij} and the pump parameters c_{2ij}, c_{1ij} and c_{0ij} which are given by the pump manufacturer. In the case $\delta_{ij} = 0$, i.e., no pump is running, the flow $q_{ij} = 0$ as check valves prevent a flow from the higher to the lower potential. Therefore, the nodes are decoupled and the head difference is directly given by $\Delta h_{\mathrm{s}ij} = h_j - h_i$. When at least one pump is running, the head increase is given by the upper relation in (7.9) which is quadratic in the flow q_{ij}. To adjust the flow to time-varying demands, either the speed ω_{ij} or the number of pumps running δ_{ij} can be altered. Although the speed could theoretically be changed continuously, in this thesis it is assumed that a small number of fixed speed levels are used which is the case in the considered scenario and most practical applications. This means that $\omega_{ij} \in \mathcal{W}_{ij}$ where $\mathcal{W}_{ij}$ is the set of discrete pump speeds. If the changes in the demands are large, several pumps are installed in parallel in a pump station. The number of pumps running in parallel δ_{ij} belongs to the set $\mathbb{Z}_{0:n_{ij}^{\mathrm{P}}}$ where n_{ij}^{P} is the number of pumps in the station.

The combinations of running pumps and speeds $(\delta_{ij}, \omega_{ij})$ are called pump configurations in the sequel. To simplify the notation, the configurations are indexed with $\kappa \in \mathbb{Z}_{1:n_{ij}^{\text{con}}}$ where the number of configurations in a pump station is called n_{ij}^{con}. By definition $\kappa = 1$ corresponds to the case where $\delta_{ij} = 0$. Note that for each configuration $\kappa \in \mathbb{Z}_{2:n_{ij}^{\text{con}}}$, δ_{ij} and ω_{ij} in (7.9) are constants and so (7.9) can be recast as

$$\Delta h_{\text{s}ij} = \begin{cases} \tilde{c}_{2ij}^{\kappa} q_{ij}^2 + \tilde{c}_{1ij}^{\kappa} q_{ij} + \tilde{c}_{0ij}^{\kappa}, & \kappa > 1 \\ h_j - h_i, & \kappa = 1 \end{cases}. \tag{7.10}$$

To put it in a nutshell, each pump configuration is modeled as a single fixed-speed pump or in other words each pump station with multiple pumps and fixed-speeds is modeled as multiple pumps with single fixed-speed. Note that only one of the configurations can be active at the same time.

A precise pump power approximation is given in [CLL16] by

$$P_{ij} = b_{0ij}^{\kappa} + b_{1ij}^{\kappa} q_{ij} + b_{2ij}^{\kappa} q_{ij}^2 + b_{3ij}^{\kappa} q_{ij}^3 \tag{7.11}$$

with the pump parameters $b_{0ij}^{\kappa}, b_{1ij}^{\kappa}, b_{2ij}^{\kappa}, b_{3ij}^{\kappa}$ which are dependent on the pump configuration and can be obtained from the pump manufacturer.

Another important component of a WDN is a reservoir where water can be stored. Analogous to (7.7), the reservoir head

$$h_{\text{r}i} = E_{\text{r}i} + x_{\text{r}i} \tag{7.12}$$

is defined by its elevation $E_{\text{r}i}$ and water level $x_{\text{r}i}$. The reservoir dynamics depend on the flow difference between ingoing flow $q_{\text{in}i}$ and outgoing flow $q_{\text{out}i}$

$$\dot{h}_{\text{r}i} = \frac{1}{S_{\text{r}i}} \left(q_{\text{in}i} - q_{\text{out}i}\right) \tag{7.13}$$

with the constant cross-sectional area $S_{\text{r}i}$.

Overall Nonlinear Model

Combining the described elements according to the given network topology, the behavior of the complete WDN can be modeled. A detailed nonlinear model of a part of the WDN in Kaiserslautern with 10 nodes has been derived. The model aggregates an area which covers over. 80 % of the overall WDN and the remaining parts are considered as demands. The nonlinear model can be described by

$$\dot{\boldsymbol{h}}_{\text{r}} = \boldsymbol{B}_{\text{flow}} \boldsymbol{q} \tag{7.14a}$$

$$\boldsymbol{A}_{\text{flow}} \boldsymbol{q} = \boldsymbol{d}_{\text{W}} \tag{7.14b}$$

$$\boldsymbol{G}(\boldsymbol{q}, \boldsymbol{\delta}, \boldsymbol{\omega}) \boldsymbol{q} + \boldsymbol{A}_{\text{flow}}^{\text{T}} \boldsymbol{h} = -\boldsymbol{C}_{\text{flow}} \boldsymbol{h}_{\text{r}}. \tag{7.14c}$$

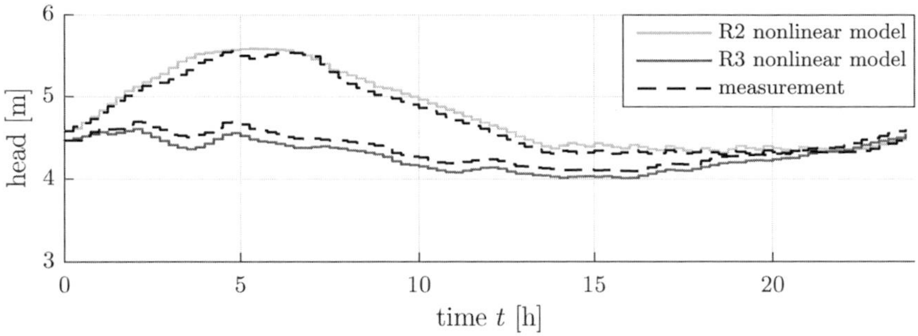

Figure 7.1: The figure shows the comparison of water levels in reservoirs from measurements and simulation using the nonlinear model.

$\boldsymbol{h}_{\mathrm{r}}$, $\boldsymbol{\delta}$, $\boldsymbol{q}$, $\boldsymbol{d}_{\mathrm{W}}$ are stacked vectors of the corresponding individual variables. The matrices $\boldsymbol{A}_{\mathsf{flow}}$, $\boldsymbol{B}_{\mathsf{flow}}$ and $\boldsymbol{C}_{\mathsf{flow}}$ describe the topology of the WDN. The mapping $\boldsymbol{G}(\boldsymbol{q}, \boldsymbol{\delta}, \boldsymbol{\omega})$ collects the nonlinear relationships, i.e., the pump characteristics and the losses. All required parameters have been taken from data sheets or identified from measurement data. The detailed system description and data cannot be provided due to a nondisclosure agreement with the WDN operator. Figure 7.1 depicts the water levels of reservoir two and three obtained from measurements and from simulation. Thereby, the power-on times and the speed of the pumps as well as the water demand has been taken from measurements and applied to the nonlinear model. One can see that the nonlinear model approximates the behavior of the plant accurately. For reservoir one, no measurements are available.

Control-oriented modeling

In this section a control-oriented model is developed for the above described system. Complex models, as derived in the previous section, are usually not feasible for controller design and simpler ones adapted to the properties of the considered WDN are required [BU94]. The structure of the control-oriented model is sketched in Figure 7.2. The model consists of $M = 3$ dominant pressure zones. The first zone has a water storage with water level $x_{\mathrm{r}1}$, an aggregated demand $d_{\mathrm{W}1}$ and an supply flow $q_{\mathrm{s}1}$ where water is pumped from deep wells. Moreover, it is connected to the second zone from which water can be pumped to zone one over a pump station with $n_{21}^{\mathrm{p}} = 2$ identical fixed-speed pumps running at $|\mathcal{W}_{21}| = 1$ speed. The number of configurations is $n_{21}^{\mathrm{con}} = 3$, as either no, one, or both pumps are running. Zone two has a water storage with water level $x_{\mathrm{r}2}$, an aggregated demand $d_{\mathrm{W}2}$. It takes the complete supply from the third system. The

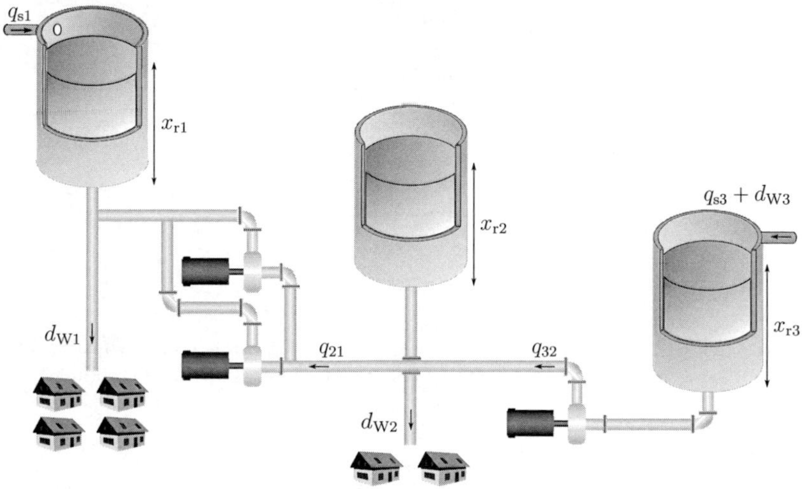

Figure 7.2: Sketch of considered water distribution network. The elevation decreases from left to right.

third zone has a water reservoir with water level x_{r3} and is a pure supply zone with supply flow q_{s3}. Furthermore, it is supplied by a spring which is considered as a negative demand in d_{W3} as there are no costs for pumping the water to the surface. From zone three one can pump water to zone two over a pump station with $n_{32}^{p} = 1$ pump which can operate at $|\mathcal{W}_{32}| = 2$ different fixed speeds. The number of pump configurations is $n_{32}^{con} = 3$ as the pump is either switched off or is running with one of the speeds.

Discretizing (7.13) with sampling interval T_s and applying (7.6), the reservoir heads h_i can be described by the discrete-time model

$$\begin{aligned} h_{1,t+1} &= h_{1,t} + \frac{T_s}{S_{r1}}(q_{s1,t} + q_{21,t} - d_{W1,t}) \\ h_{2,t+1} &= h_{2,t} + \frac{T_s}{S_{r2}}(q_{32,t} - q_{21,t} - d_{W2,t}) \\ h_{3,t+1} &= h_{3,t} + \frac{T_s}{S_{r3}}(q_{s3,t} - q_{32,t} - d_{W3,t}). \end{aligned} \tag{7.15}$$

In order to simplify the notation, the index r in h_{ri} is dropped from now on. The reservoir heads have to be operated in the bounds

$$h_i^{\min} \leq h_i \leq h_i^{\max} \tag{7.16}$$

for all $i \in \mathbb{Z}_{1:3}$. $h_i^{\min}$ and $h_i^{\max}$ are minimal and maximum head in reservoir i. The flows in the system are constrained to

$$q_{si}^{\min} \leq q_{si} \leq q_{si}^{\max} \quad i \in \{1,3\}, \tag{7.17}$$
$$q_{21}^{\min} \leq q_{21} \leq q_{21}^{\max} \tag{7.18}$$
$$q_{32}^{\min} \leq q_{32} \leq q_{32}^{\max}. \tag{7.19}$$

To account for the losses in the pipes of the zones, they are lumped into the links between the pressure zones. Due to the almost quadratic form of (7.8), the approximation

$$\Delta h_{\mathrm{p}ij} = \tilde{R}_{ij} q_{ij}^2 \tag{7.20}$$

can be applied. $\tilde{R}_{ij}$ can be estimated via measurements. The losses are then included into the pump characteristic (7.10) which leads to

$$\Delta h_{ij} = \begin{cases} \left(\tilde{c}_{2ij}^{\kappa} - \tilde{R}_{ij}\right) q_{ij}^2 + \tilde{c}_{1ij}^{\kappa} q_{ij} + \tilde{c}_{0ij}^{\kappa}, & \kappa > 1 \\ h_j - h_i, & \kappa = 1. \end{cases} \tag{7.21}$$

Integrating (7.21) into an optimization problem would lead to a nonlinear mixed-integer program. As such problems can be hard to solve, an approximation is introduced. Thereby, the pump characteristic of each pump configuration is approximated by a polytope around the nominal operation point. The procedure for determining the approximation is described in the following and shown in Figure 7.3. In a first step the range $(h_i, h_j) \in \mathcal{H}_{ij} \subseteq \mathbb{R}^2$ in which $\Delta h_{ij} = h_j - h_i$ can lie is determined. This range is bounded as the water levels in the reservoirs are bounded. Then a straight line is determined through the intersection points of $\mathcal{H}_{ij}$ with the pump characteristic curve (lower approximation). A predefined number of lines which are tangentially to the pump characteristic curve build the upper approximation. The intersection of the halfspaces spanned by these lines defines the polytope $\tilde{\mathcal{P}}_{ij}^{\kappa} = \left\{\boldsymbol{x} \in \mathbb{R}^3 | \boldsymbol{H}_{ij}^{\kappa} \boldsymbol{x} \leq \boldsymbol{f}_{ij}^{\kappa}\right\}$ for a configuration $\kappa \in \mathbb{Z}_{2:n_{ij}^{\mathrm{con}}}$.

As mentioned before, when the pumps in the stations are switched off, check valves prevent a flow from the higher potential to the lower one. In this case the nodes are decoupled, $q_{ij} = 0$ and the head drop is directly given by $\Delta h_{ij} = h_j - h_i$. Thus, heads and flow are in the polytope $\tilde{\mathcal{P}}_{ij}^{1} = \{x_1, x_2, x_3 \in \mathbb{R} | x_2 - x_1 \in \mathcal{H}_{ij} \wedge x_3 = 0\}$.

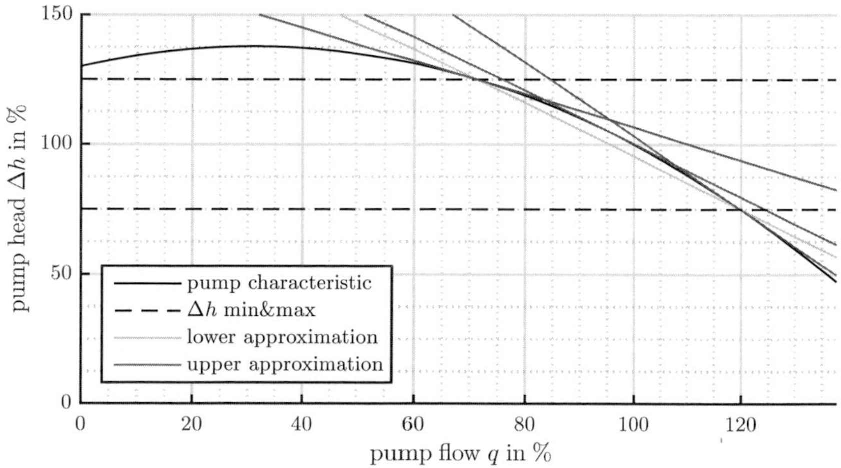

Figure 7.3: Polytopic approximation of pump characteristic around nominal operation point

Considering all configurations in a pump station, one can define the union of the polytopes $\mathcal{P}_{ij} = \bigcup_{\kappa \in \mathbb{Z}_{1:n_{ij}^{\mathrm{con}}}} \tilde{\mathcal{P}}_{ij}^{\kappa}$. To ensure that (h_i, h_j, q_{ij}) satisfy the pump characteristic approximately, the constraint

$$(h_i, h_j, q_{ij}) \in \mathcal{P}_{ij} \tag{7.22}$$

can be included into the controller optimization problem as shown later. Note that the constraint is non-convex but can be integrated in an Mixed-Integer Quadratic Program (MIQP) as described in the sequel.

The WDN described by (7.15) can be recast in the form (7.1). Thereby, a subsystem corresponds to a pressure zone and $\boldsymbol{x}_i = h_i$ for all $i \in \mathbb{Z}_{1:3}$, $\boldsymbol{u}_1 = \begin{pmatrix} q_{\mathrm{s1}} & q_{21} \end{pmatrix}^{\mathrm{T}}$, $\boldsymbol{u}_2 = q_{32}$ and $\boldsymbol{u}_3 = q_{\mathrm{s3}}$. The matrices are $\boldsymbol{A}_{ii,t} = 1$ for all $i \in \mathbb{Z}_{1:M}$, $\boldsymbol{B}_{11,t} = \frac{T_{\mathrm{s}}}{S_{\mathrm{r1}}} \begin{pmatrix} 1 & 1 \end{pmatrix}$, $\boldsymbol{B}_{22,t} = \frac{T_{\mathrm{s}}}{S_{\mathrm{r2}}}$ and $\boldsymbol{B}_{21,t} = -\frac{T_{\mathrm{s}}}{S_{\mathrm{r2}}}$ $\boldsymbol{B}_{33,t} = \frac{T_{\mathrm{s}}}{S_{\mathrm{r3}}}$ and $\boldsymbol{B}_{32,t} = -\frac{T_{\mathrm{s}}}{S_{\mathrm{r3}}}$ for all $t \in \mathbb{N}$. The exogenous input $\boldsymbol{d}_i = -\frac{T_{\mathrm{s}}}{S_{\mathrm{r1}}} d_{\mathrm{W}i}$ is assumed to be periodic with period $P \in \mathbb{N}$, i.e., $\boldsymbol{d}_{i,t} = \boldsymbol{d}_{i,t+P}$ for all $t \in \mathbb{N}$. For a water system this assumption approximately holds with a period of one day as the water demands $d_{\mathrm{W}i}$ do not change significantly from day to day. Due to this the system is LPTV. The coupling structure in the dynamics is $\mathcal{N}_1 = \emptyset$, $\mathcal{N}_2 = \{1\}$ and $\mathcal{N}_3 = \{2\}$. Subsystems with this coupling structure are also called cascaded subsystems. Each subsystem optimizes over its own input, i.e., $\mathcal{N}_i^{\mathrm{U}} = \{i\}$ for all $i \in \mathbb{Z}_{1:3}$. The set of states that are directly affected are $\mathcal{N}_1^{\mathrm{X}} = \{1, 2\}$, $\mathcal{N}_2^{\mathrm{X}} = \{2, 3\}$ and $\mathcal{N}_3^{\mathrm{X}} = \{3\}$.

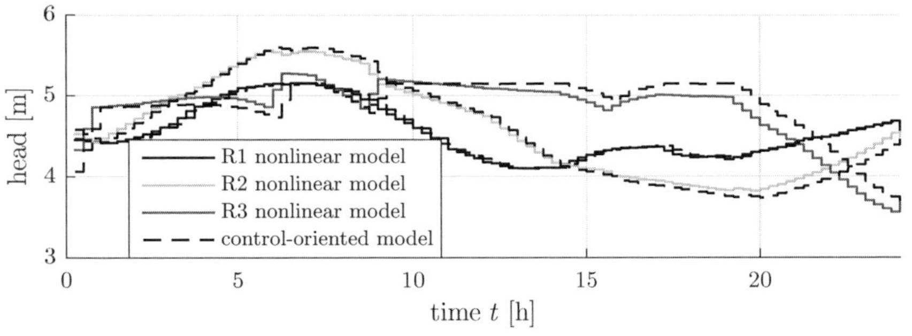

Figure 7.4: The figure shows the comparison of the water levels in the reservoirs computed with nonlinear and control-oriented model.

Moreover, constraints are (7.16) and (7.17) are written as

$$\mathcal{X}_{i,t} = \left\{\boldsymbol{x}_i \in \mathbb{R} | \boldsymbol{x}_i^{\min} \leq \boldsymbol{x}_i \leq \boldsymbol{x}_i^{\max}\right\}, \mathcal{U}_{i,t} = \left\{\boldsymbol{u}_i \in \mathbb{R}^{m_i} | \boldsymbol{u}_i^{\min} \leq \boldsymbol{u}_i \leq \boldsymbol{u}_i^{\max}\right\} \tag{7.23}$$

for all $i \in \mathbb{Z}_{1:3}$ and $t \in \mathbb{N}$ with $\boldsymbol{x}_i^{\min} = h_i^{\min}$ and $\boldsymbol{x}_i^{\max} = h_i^{\max}$ for all $i \in \mathbb{Z}_{1:3}$ as well as $\boldsymbol{u}_1^{\min} = \left(q_{\mathrm{s}1}^{\min}, q_{21}^{\min}\right)^{\mathrm{T}}$, $\boldsymbol{u}_1^{\max} = \left(q_{\mathrm{s}1}^{\max}, q_{21}^{\max}\right)^{\mathrm{T}}$, $\boldsymbol{u}_2^{\min} = q_{32}^{\min}$, $\boldsymbol{u}_2^{\max} = q_{32}^{\max}$, $\boldsymbol{u}_3^{\min} = q_{\mathrm{s}3}^{\min}$, $\boldsymbol{u}_3^{\max} = q_{\mathrm{s}3}^{\max}$, $m_1 = 2$, $m_2 = m_3 = 1$. Additionally, the constraints (7.22) can be recast as (7.3), i.e., $\mathcal{Z}_{1,t}^{\mathrm{C}} = \mathcal{P}_{12}$ and $\mathcal{Z}_{2,t}^{\mathrm{C}} = \mathcal{P}_{23}$ for all $t \in \mathbb{N}$. Consequently there are $n_{\mathrm{c}} = 2$ coupling constraints and their structure is $\mathcal{N}_1^{\mathrm{CX}} = \{1, 2\}$, $\mathcal{N}_1^{\mathrm{CU}} = \{1\}$, $\mathcal{N}_2^{\mathrm{CX}} = \{2, 3\}$, $\mathcal{N}_2^{\mathrm{CU}} = \{2\}$.

A comparison between the behavior of the control-oriented and the nonlinear model is shown in Figure 7.4. It shows the water levels in the reservoirs when applying the inputs obtained from an optimization problem subject to the control-oriented model to the detailed nonlinear model in an open-loop fashion. The small deviation shows that the control-oriented model is a good approximation for the nonlinear model and thus for the behavior of the plant.

Next, the cost function is discussed. The electricity cost is determined by multiplying the time-varying price $c_{\mathrm{E},t}$ for energy with the electrical power consumed by the pump over the sampling interval. The power relation (7.11) for the pumps connecting pressure zones can be approximated in the operation area by a piecewise affine function, i.e.,

$$P_{ij}^{\kappa}(q_{ij}) = \tilde{b}_{1ij}^{\kappa} q_{ij} + \tilde{b}_{0ij}^{\kappa} \text{ for } q_{ij}^{\min,\kappa} \leq q_{ij} \leq q_{ij}^{\max,\kappa} \tag{7.24}$$

where $\tilde{b}_{1ij}^{\kappa}, \tilde{b}_{2ij}^{\kappa} \in \mathbb{R}$ are constant and κ indexes again the configuration. $q_{ij}^{\min,\kappa}$ and $q_{ij}^{\max,\kappa}$ are minimal and maximum flow for a configuration. For the case that all pumps are switched off, $P_{ij}^{1}(q_{ij}) = 0$.

The flow q_{si} aggregates the supply flows of multiple pumps in zone i pumping water from deep wells. A lower-level controller schedules the pumps in the deep wells to provide the required flow q_{si}. It has to consider legal and technical limitations but is not further discussed here. The power consumption dependent on the flow q_{si} has been determined from measurement data and can be approximated by the quadratic function

$$P_{\mathrm{si}}(q_{\mathrm{si}}) = a_{2i}q_{\mathrm{si}}^2 + a_{1i}q_{\mathrm{si}} + a_{0i}. \tag{7.25}$$

Thus, the total cost for electricity can be recast in the stage cost as

$$l_{i,t}^{\mathrm{u}}(\boldsymbol{u}_{i,t}) = \boldsymbol{u}_{i,t}^{\mathrm{T}}\boldsymbol{R}_{i,t}\boldsymbol{u}_{i,t} + \boldsymbol{r}_{i,t}^{\kappa,\mathrm{T}}\boldsymbol{u}_{i,t} + r_{i,t}^{\kappa} \tag{7.26}$$

where $\boldsymbol{R}_{i,t}$, $\boldsymbol{r}_{i,t}^{\kappa}$, $r_{i,t}^{\kappa}$ are weights combining the electricity price $c_{\mathrm{E},t}$ and the pump power characteristics (7.24) and (7.25). They are periodic with period P (due to the electricity price), i.e., $\boldsymbol{R}_{i,t} = \boldsymbol{R}_{i,t+P}$, $\boldsymbol{r}_{i,t}^{\kappa} = \boldsymbol{r}_{i,t+P}^{\kappa}$ and $r_{i,t}^{\kappa} = r_{i,t+P}^{\kappa}$ for all $t \in \mathbb{N}$.

Moreover, the cost term

$$l_i^{\mathrm{x}}(\boldsymbol{x}_i) = (\boldsymbol{x}_i - \boldsymbol{x}_{i,\mathrm{ref}})^{\mathrm{T}}\boldsymbol{Q}_i(\boldsymbol{x}_i - \boldsymbol{x}_{i,\mathrm{ref}}) \tag{7.27}$$

is included where $\boldsymbol{x}_{\mathrm{ref},i}$ is a constant reference for the state and $\boldsymbol{Q}_i$ is a positive definite weighting matrices. l_i^{x} penalizes the deviation of the actual state to the reference. The term can be used so that the water level in the reservoirs does not become too low which is important due to safety requirements in cases of blackouts in the electrical power system or fire.

The total stage cost is given by

$$l_{i,t}(\boldsymbol{x}_{i,t}, \boldsymbol{u}_{i,t}) = l_{i,t}^{\mathrm{u}}(\boldsymbol{u}_{i,t}) + l_i^{\mathrm{x}}(\boldsymbol{x}_{i,t}). \tag{7.28}$$

Together with the cost (7.28), optimization problem (7.5) can be recast as an MIQP for which efficient solvers, such as CPLEX® [IBM14], exist. The integer variables stem from the switching of the pumps, i.e., correspond to the possible configurations in the pump stations.

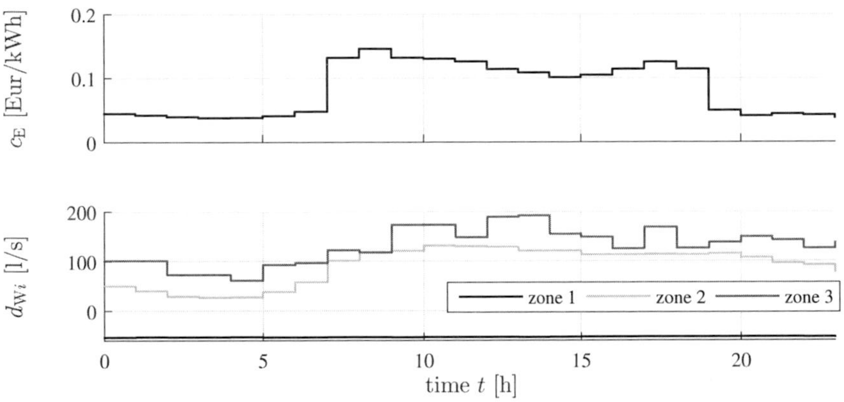

Figure 7.5: Price for electrical power $c_{E,t}$ and water demands $d_{Wi,t}$ over one day

7.2.2 Simulation Example

Simulation Set-up

In this section the DMPC scheme presented in Section 7.1 is applied to the considered WDN. The sampling interval is chosen to $T_s = 1\,\text{h}$. The electricity price and water demands d_{Wi} are shown in Figure 7.5 for one day. As mentioned before they are assumed to be periodic with period of one day, i.e., $P = 24$. The prediction horizon is set to $N = P = 24$.

It is worth mentioning that the first zone could be operated independently as the deep wells could cover the demand in this zone. However, as the first system has a higher elevation, the costs for pumping water there are higher compared to them in zone three. Hence, it might be beneficial that zone three supplies water to zone one.

As mentioned before, for the initialization of the distributed algorithm at time $t = 0$, initially feasible input trajectories must be determined. To prevent a centralized initialization, the WDN is divided into self-sustaining areas. For this purpose, the pump station between subsystem one and two is shut off. Thus, subsystem one solves a decentralized optimization problem where the connection between the first and the second subsystem is set to zero, i.e., $q_{21} = 0$. Additionally, subsystem two and three solve jointly an optimization problem with $q_{21} = 0$. The obtained input trajectories are used to start the algorithm. This approach can also be applied for larger systems.

As pointed out in Remark 7.1.3 in case of a failure of one of the controllers, it might be possible to keep up the control of the other subsystems. In the considered WDN, in case of a failure in zone one, one can still control zone two and three and vice versa as those zones can be operated independently. Due to similar reasons as outlined in the initialization approach, this idea can also be transferred to larger systems.

Simulation Results for Control-oriented Model

The simulation results of the DMPC are compared to a Centralized Model Predictive Control (CMPC) scheme. The controller data for centralized and distributed control are identical. The state and input trajectory are plotted over four periods in Figure 7.6. In the simulation the optimization problem is feasible at all time (recursively feasibility) and it can be observed that the constraints are respected and the trajectories of both controllers converge to the optimal sequence computed by solving problem (2.24) offline. The controllers exploit the storage capacity of the reservoirs by filling them in the morning time when the energy price is low and emptying them when the price is high (cf. Figure 7.5). Furthermore, water is pumped from zone three over zone two to zone one as the cost for the water supply in zone three is lower as in zone one due to the lower elevation.

Considering Figure 7.7 one can see that the pumps are operated close to their actual flow-head characteristic curve and the presented polytopic approximation is sufficiently accurate. The power (7.11) is also approximated satisfactorily by (7.24). Moreover, the pump connecting subsystems two and three is operated in both fixed speeds.

The average times required to solve the optimization problems are $t_{\mathrm{DMPC}} = 0.2753\,\mathrm{s}$ and $t_{\mathrm{CMPC}} = 0.4003\,\mathrm{s}$. This shows that the controller is real-time applicable and that for the considered scenario it is in average faster to solve several smaller problems in parallel than one large centralized one. In Figure 7.6 one can also see the number iterations at each time step. At the beginning the controllers can improve the cost with the new calculated inputs. When the DMPC trajectories approach the optimal sequences, the improvement is too small and thus no more iterations are required.

The influence of the threshold parameter η_i on the communication effort and the performance can be seen in Table 7.2. Thereby, the closed-loop costs are determined by $J_{\mathrm{cl}} = \sum_{t=0}^{T_{\mathrm{sim}}} l_t\left(\boldsymbol{x}_t, \boldsymbol{u}_t\right)$. One can see that with increasing threshold, the number of iterations increases and the cost is decreases. Comparing the cost with the MPC controller, one can see that the DMPC is performing slightly worse due to the distributed decision making and the incomplete knowledge in the subsystems.

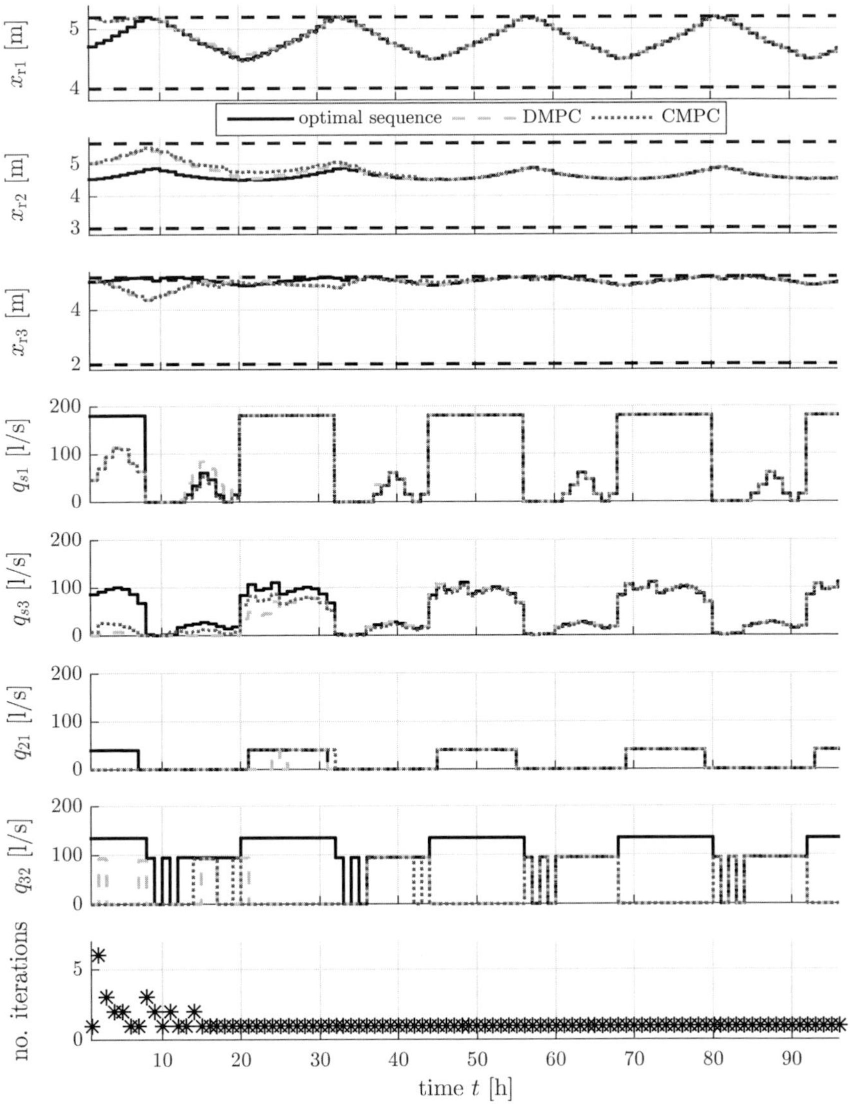

Figure 7.6: The figure shows the simulation results over four periods. The black dashed lines show the water level constraints.

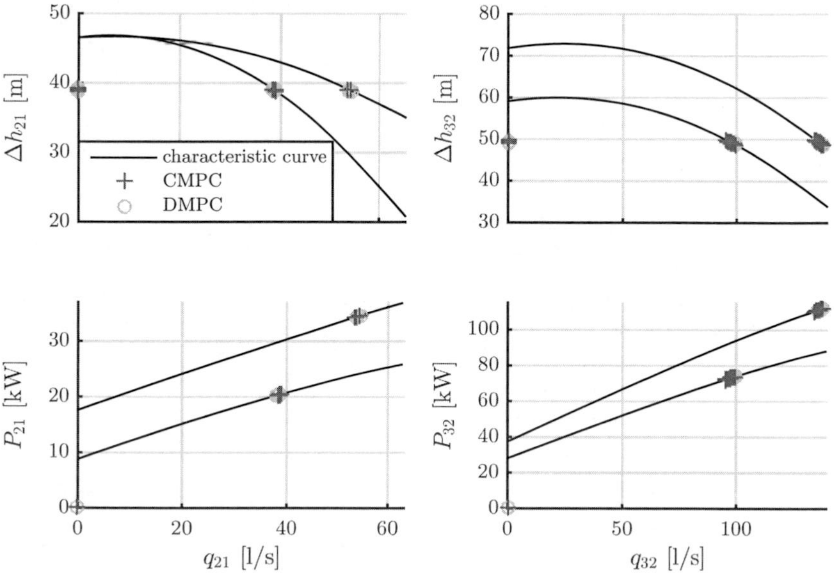

Figure 7.7: Pump characteristic curves and operating points in simulation

Table 7.2: Influence of the threshold parameter on the performance and communication effort

	DMPC				CMPC
η_i	-1	-0.5	-0.2	-0.1	-
J_{cl}	970.45	970.46	969.50	966.52	961.45
no. iterations	96	100	103	112	-

Simulation Results for Nonlinear Model

Next, the case is discussed when the controller is applied to the nonlinear model (7.14). The control-oriented model is still used in the controller. As discussed in Remark 7.1.2, due to the deviation between control-oriented and nonlinear model, feasibility can be lost. Furthermore, it was mentioned that a robust control design due to non-convexity is challenging. For this purpose a "practical" approach is presented in the sequel. To avoid violations of the hard constraints, soft-constraints for the reservoir heads are introduced. They are of the form

$$h^{\min}_{\mathrm{soft}i} - e^{\min}_{i,t} \leq h_{i,t} \leq h^{\max}_{\mathrm{soft}i} + e^{\max}_{i,t} \tag{7.29}$$

$$e^{\min}_{i,t}, e^{\max}_{i,t} \geq 0 \qquad \forall i \in \mathbb{Z}_{1:M} \tag{7.30}$$

with the tightened constraints $h^{\max}_{\mathrm{soft}i} = h^{\max}_i - \Delta h_{\mathrm{soft}i}$ and $h^{\min}_{\mathrm{soft}i} = h^{\min}_i + \Delta h_{\mathrm{soft}i}$ where $\Delta h_{\mathrm{soft}i} \geq 0$ are constants. $e^{\min}_{i,t}$ and $e^{\max}_{i,t}$ are additional variables which are called soft constraint variables in the sequel. Subsystem $\mathscr{S}_i$ has to include the soft constraint variables in the optimization (7.5) $\forall j \in \mathcal{N}^{\mathrm{X}}_i$ and $\mathscr{S}_{i^*}$ needs to transmit the solution to $\mathscr{S}_j$ with $j \in \left\{s \in \mathbb{Z}_{1:M} | i^* \in \mathcal{N}^{\mathrm{X}}_s\right\}$ in step 5 of Algorithm 4.

Furthermore, the terminal constraint (7.5c) is removed and and the cost (7.5a) is replaced for all $i \in \mathbb{Z}_{1:M}$ by

$$\sum_{j\in\mathcal{N}^{\mathrm{X}}_i} V^{\mathrm{f}}_{j,t}\left(\boldsymbol{x}_{j,t+N|t}\right) + \sum_{r=t}^{t+N-1} l^{\mathrm{e}}_j\left(\boldsymbol{e}_{j,r|t}\right) + \sum_{j\in\mathcal{N}^{\mathrm{X}}_i\cup\mathcal{N}^{\mathrm{U}}_i} \sum_{r=t}^{t+N-1} l_{j,r}\left(\boldsymbol{x}_{j,r|t}, \boldsymbol{u}_{j,r|t}\right) \tag{7.31}$$

where for $i \in \mathbb{Z}_{1:M}$:

$$V^{\mathrm{f}}_{i,t}\left(\boldsymbol{x}_{i,t+N|t}\right) = \left(\boldsymbol{x}_{i,t+N|t} - \boldsymbol{x}^{\mathrm{P}}_{i,[t+N]_P}\right)^{\mathrm{T}} \boldsymbol{P}_i \left(\boldsymbol{x}_{i,t+N|t} - \boldsymbol{x}^{\mathrm{P}}_{i,[t+N]_P}\right) \tag{7.32}$$

$$l^{\mathrm{e}}_i\left(\boldsymbol{e}_i\right) = \boldsymbol{e}^{\mathrm{T}}_i \boldsymbol{Q}_{\mathrm{e},i} \boldsymbol{e}_i \tag{7.33}$$

where $\boldsymbol{e}_i = \left(e^{\min}_i, e^{\max}_i\right)^{\mathrm{T}}$, $\boldsymbol{Q}_{\mathrm{e},i} \in \mathbb{R}^{2\times2}$ with $\boldsymbol{Q}_{\mathrm{e},i} \succ \boldsymbol{0}$ and $\boldsymbol{P}_i \in \mathbb{R}_{>0}$ are weighting matrices. Compared to (7.5a) the cost term l^{e}_i for the soft constraints and the terminal cost $V^{\mathrm{f}}_{i,t}$ is added. The first term forces the controller to avoid violating the soft constraints as this causes additional costs. The second term ensures that the last state $\boldsymbol{x}_{i,t+N|t}$ of the predicted sequence is close to corresponding state of the periodic sequence $\boldsymbol{x}^{\mathrm{P}}_{i,[t+N]_P}$. Thereby, it approximates the removed terminal constraint. To obtain a periodic sequence which can be exploited in the terminal cost $V^{\mathrm{f}}_{i,t}$, optimization problem (2.24) is solved using the control-oriented model. Notably, the resulting sequence is not optimal for the nonlinear model. Due to the disturbance, the cost for the first iteration cannot be determined anymore by shifting the previous solution. Therefore, in Algorithm 4 the cost in the first iteration is set to infinity. This implies that there is at least two iterations per time step.

The simulation results are shown in Figure 7.8. The basic behavior is similar to the previous case. The control law remains feasible and it can be observed that the constraints are satisfied at all times. Furthermore, the state and input trajectories approach the periodic sequence and remain close to it but there is a small deviation. This can be explained by the fact that the offline determined periodic sequence is not optimal for the nonlinear model and by the difference between actual and predicted evolution of the state. Moreover, one can see that the controller iterates not only at the beginning because of the deviation between control-oriented and nonlinear model which forces the controller to adjust the input even when it operates close to the periodic sequence.

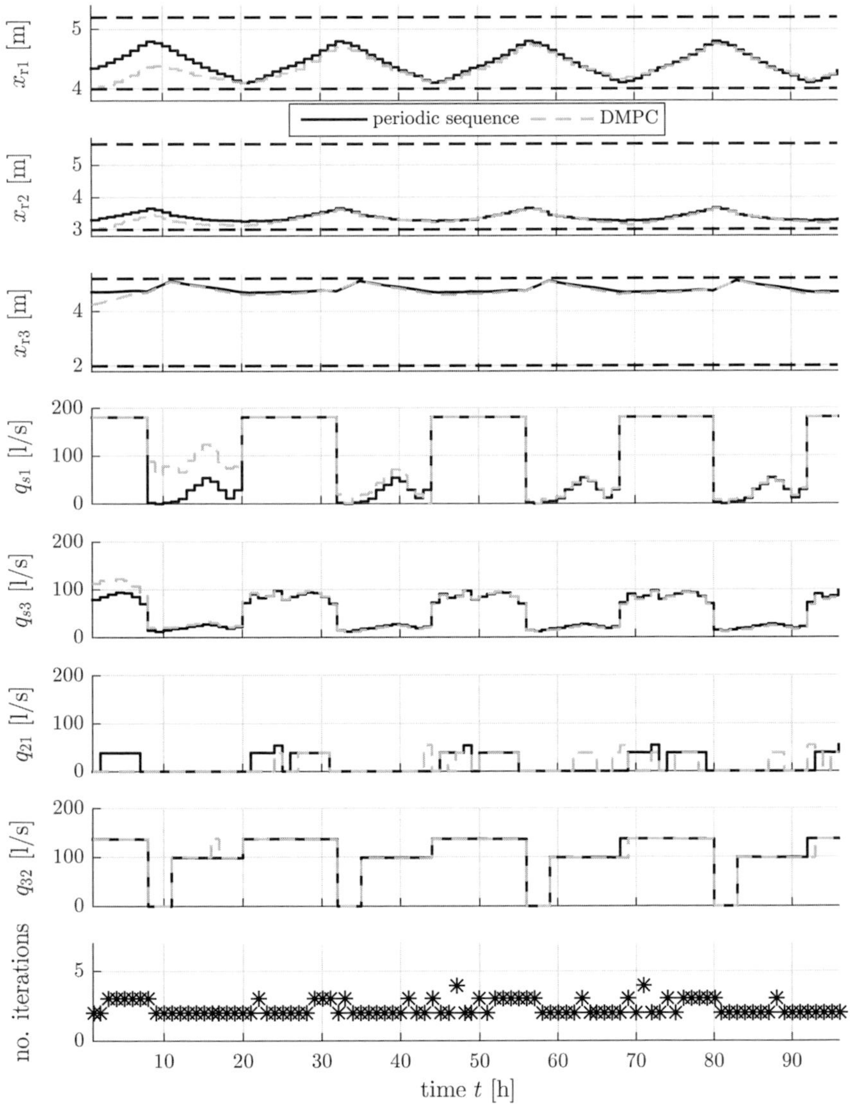

Figure 7.8: Simulation results over four periods when the DMPC with control-oriented model is applied to the nonlinear model

7.3 Application to Voltage Control in MV Grids

Problem Set-up

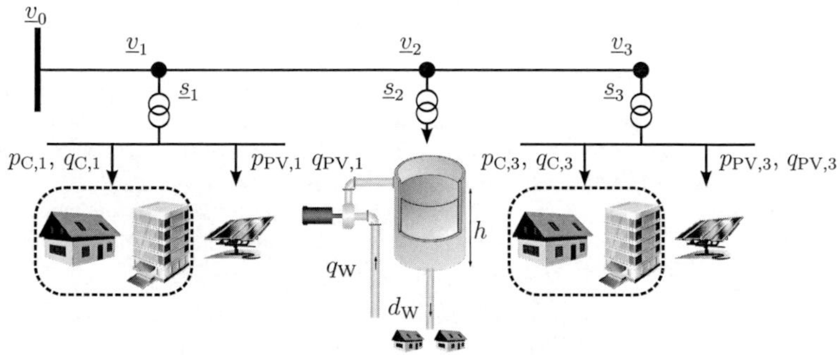

Figure 7.9: The figure shows the structure of the considered radial MV power grid. $\underline{v}_i$ are the complex voltages and $\underline{s}_i$ the apparent power in the nodes $i \in \{1, 2, 3\}$. The consumers in node 1 and 3 draw the active and reactive powers $p_{C,i}$ and $q_{C,i}$ with $i \in \{1, 3\}$ form the grid, respectively. The PV plants inject the active power $p_{PV,i}$ and reactive power $q_{PV,i}$ with $i \in \{1, 3\}$ into the grid. In node 2 the pumps of a supply zone of a WDN are connected. The pump flows are aggregated to the flow q_W which fills a reservoir with head h. The uncontrollable water demand d_W has to be satisfied.

In this section a radial medium voltage grid with $M = 3$ nodes as shown in Figure 7.9 is considered. The set-up is derived from a real-world example in the city of Kaiserslautern. In the first and the third node there are aggregated consumers (households and companies) which draw active and reactive power from the grid. The demand from those consumers is assumed to be uncontrollable. Moreover, solar plants, which are connected to the grid via converters, are located there and inject active as well as reactive power. In the second node a pump station of a WDNs is installed which is fed by the power grid. The major purpose of the pump station is to supply water to the consumers of the WDN. The WDN comprises a water reservoir in which water can be stored.

In the considered set-up it is investigated how the pumping in the WDNs can be incorporated into the control of the voltages in the grid. As the system contains storages in the form of water reservoirs there is some degree of freedom to coordinate the pumping of water with the energy sources and remaining consumers in order to keep the voltage in its limitations as well as to satisfy simultaneously the water demand in the grid. The combined consideration makes sense from a practical point of view since both infrastructure systems belong to the same operation and so there is an interest in developing a holistic solution. Note that in today's power grids this potential is usually unused. Additionally, the two PV arrays in the grid, which are controllable, can support the voltage control by supplying reactive power to the grid. Although it would be technically feasible, a reduction of the feed-in power from the PV array is not considered, as this is usually undesired from an economic point of view.

In both infrastructure systems, the power grid and the WDN, there are uncontrollable demands which are almost periodic from day to day. Hence, the combined system is modeled as an LPTV system which can be exploited to analyze the theoretical properties in the system and to design a recursive feasible controller. It is subject to non-convex constraints since the WDN contains pumps which have a non-continuous characteristic. The system can be decomposed into subsystems which correspond to the nodes and have decoupled dynamics but are subject to a common voltage constraint.

The control objective is to keep the voltages within its desired boundaries. It is desired that the controllers is implementable in a distributed fashion due to the increased robustness and reduced complexity properties. For this purpose Algorithm 4 presented in Section 7.1 is applied.

Notably, although MPC and DMPC are well suited for voltage control in Low Voltage (LV) and MV grids due to their ability to systematically handle constraints and systems with multiple inputs and outputs, the available results in the literature are rare. For LV grids, DMPC methods with a focus on data privacy are introduced in [BFB$^+$16, BBL19]. Hierarchical solutions for MV grids are proposed in [FGM$^+$15, BC18] where centralized model predictive controllers are applied at the higher level and proportional-integral controllers track their references at lower level. The joined coordination of two infrastructure systems utilizing DMPC has been studied for gas and electricity grids in [ANAS10] where a decomposition-based approach is applied. Moreover, for the considered set-up also the decomposition-based approaches proposed in [BGB12] which considers systems with input that belong to discrete sets could be applied. However, the high communication effort of these approaches makes them unattractive for the considered application.

7.3.1 Modeling of MV Grids

Power Grid

To quantify the effect of the powers on the voltages and currents in the grid, a power flow problem has to be solved. The problem is shortly introduced for the considered radial set-up in the sequel. Then, a linear approximation for the power flow in the grid is briefly reviewed. More information can be found in [BBL19]. After that, the remaining parts of the grid are modeled. In order to avoid confusion between the current and the index of the subsystem, the index of the subsystem is called ν in this section. All powers drawn from/fed into the grid are by definition negative/positive.

Based on Z- and P-load models, the currents $\underline{i}_\nu$ and voltages $\underline{v}_\nu$ in the node ν of the radial grid can be described by

$$\underline{i}_\nu = -\frac{\underline{v}_\nu}{\underline{z}_\nu} + \frac{\overline{\underline{s}}_\nu}{\overline{\underline{u}}_\nu}, \tag{7.34a}$$

$$\underline{v}_\nu = \boldsymbol{W}_\nu \underline{\boldsymbol{Z}} \underline{\boldsymbol{i}} + \underline{v}_0 \tag{7.34b}$$

for all $\nu \in \mathbb{Z}_{1:M}$. $\underline{z}_\nu$ is the load impedance and $\underline{s}_\nu$ the apparent power in the node. $\underline{v}_0$ is the voltage in the slack bus and $\underline{\boldsymbol{i}} = \mathrm{col}_{\nu\in\mathbb{Z}_{1:M}}(\underline{i}_\nu)$. $\underline{\boldsymbol{Z}}$ is the unique constant impedance matrix of the grid [BCCC11]. The transformation matrix $\boldsymbol{W}_\nu$ maps the corresponding voltage drops across the power line from the slack bus to node ν. For given values $\underline{v}_0$, $\underline{z}_\nu$ and $\underline{s}_\nu$, the nonlinear power flow equations (7.34) can be solved for the voltages $\underline{v}_\nu$ and currents $\underline{i}_\nu$ for all ν. For this purpose, for instance, the Forward Backward Sweep (FBS) method can be applied.

For the controller implementation, a linear relation between current, voltage and power is appealing. Substituting (7.34b) into (7.34a) for all $\nu \in \mathbb{Z}_{1:M}$, linearizing the resulting equations around an operating point and separating them into real and imaginary part one obtains the linear relation

$$\boldsymbol{0} = \boldsymbol{A}_{\mathrm{PF}} \cdot \boldsymbol{i} + \boldsymbol{B}_{\mathrm{PF}} \cdot \boldsymbol{s} + \boldsymbol{c}_{\mathrm{PF}} \tag{7.35a}$$

$$\boldsymbol{v} = \boldsymbol{C}_{\mathrm{PF}} \cdot \boldsymbol{i} + \boldsymbol{v}_0, \tag{7.35b}$$

$$\text{with } \boldsymbol{v}_0 = \begin{pmatrix} \boldsymbol{1} \cdot \Re\{\underline{v}_0\} \\ \boldsymbol{1} \cdot \Im\{\underline{v}_0\} \end{pmatrix}, \boldsymbol{v} = \begin{bmatrix} \boldsymbol{v}^{\mathrm{Re}} \\ \boldsymbol{v}^{\mathrm{Im}} \end{bmatrix}, \boldsymbol{s} = \begin{bmatrix} \boldsymbol{p} \\ \boldsymbol{q} \end{bmatrix}, \boldsymbol{i} = \begin{bmatrix} \boldsymbol{i}^{\mathrm{Re}} \\ \boldsymbol{i}^{\mathrm{Im}} \end{bmatrix}$$

where $\boldsymbol{i}^{\mathrm{Re}} = \mathrm{col}_{\nu\in\mathbb{Z}_{1:M}}(\Re(\underline{i}_\nu))$ and $\boldsymbol{i}^{\mathrm{Im}} = \mathrm{col}_{\nu\in\mathbb{Z}_{1:M}}(\Im(\underline{i}_\nu))$ as well as $\boldsymbol{v}^{\mathrm{Re}} = \mathrm{col}_{\nu\in\mathbb{Z}_{1:M}}(\Re(\underline{v}_\nu))$ and $\boldsymbol{v}^{\mathrm{Im}} = \mathrm{col}_{\nu\in\mathbb{Z}_{1:M}}(\Im(\underline{v}_\nu))$ are the stacked real and imaginary parts of the node currents as well as voltages, respectively. $\boldsymbol{p} = \mathrm{col}_{\nu\in\mathbb{Z}_{1:M}}(\Re(\underline{s}_\nu))$ and $\boldsymbol{q} = \mathrm{col}_{\nu\in\mathbb{Z}_{1:M}}(\Im(\underline{s}_\nu))$ are the stacked active and reactive node powers. For the derivation and the computation of the constant matrices $\boldsymbol{A}_{\mathrm{PF}} \in \mathbb{R}^{2M\times 2M}$, $\boldsymbol{B}_{\mathrm{PF}} \in \mathbb{R}^{2M\times 2M}$ and $\boldsymbol{C}_{\mathrm{PF}} \in \mathbb{R}^{2M\times 2M}$ please refer to [BBL19]. Notably, the matrix $\boldsymbol{A}_{\mathrm{PF}}$ has full rank. The capacitance of the cables is considered in the additional term $\boldsymbol{c}_{\mathrm{PF}} \in \mathbb{R}^{2M\times 1}$. Moreover, the local stacked voltages $\boldsymbol{v}_\nu^{\mathrm{T}} = [\Re\{\underline{v}_\nu\}\ \Im\{\underline{v}_\nu\}]$, currents $\boldsymbol{i}_\nu^{\mathrm{T}} = [\Re\{\underline{i}_\nu\}\ \Im\{\underline{i}_\nu\}]$ and apparent powers $\boldsymbol{s}_\nu^{\mathrm{T}} = [\Re\{\underline{s}_\nu\}\ \Im\{\underline{s}_\nu\}] = [p_\nu\ q_\nu]$ are introduced.

Remark 7.3.1. *In the derived approximation, the nonlinear equations are linearized around a time-invariant operating point of the grid. As outlined in [BBL19], the presented linear power flow approximation can be adapted to time-varying operating points* *and thus its accuracy can be improved. Another option is the linearization around a periodic trajectory. The resulting matrices could be included into the outlined controller design in a straightforward manner.*

To allow a safe operation of the power grid, the nodal voltage $\boldsymbol{v}_\nu$ is desired to lie in the set

$$\tilde{\mathcal{V}}_\nu = \left\{ \boldsymbol{v}_\nu \in \mathbb{R}^2 | v_{\text{min}}^2 \leq \|\boldsymbol{v}_\nu\|^2 \leq v_{\text{max}}^2 \right\}. \tag{7.36}$$

Note that the set is non-convex. v_{min} and v_{max} are the minimum and maximum root mean squares of the voltage, respectively. To obtain a computationally more attractive set, first, the set

$$\tilde{\mathcal{V}}_\nu^{\text{a}} = \left\{ \begin{pmatrix} v_\nu^1 \\ v_\nu^2 \end{pmatrix} \in \mathbb{R}^2 | \left\| \begin{pmatrix} v_\nu^1 \\ v_\nu^2 \end{pmatrix} \right\|^2 \leq v_{\text{max}}^2 \wedge v_\nu^1 \geq v_{\text{min}} \right\}. \tag{7.37}$$

is introduced. Here the non-convex lower boundary is replaced by a linear inequality. Note that the approximation is sufficiently accurate as the voltage angle does not vary strongly for the considered type of MV grids. Then, a polytopic inner-approximation of the set $\tilde{\mathcal{V}}_\nu^{\text{a}}$ is introduced. It is of the form

$$\mathcal{V}_\nu = \left\{ \boldsymbol{v}_\nu \in \mathbb{R}^2 | \boldsymbol{H}_\nu^{\text{v}} \boldsymbol{v}_\nu \leq \boldsymbol{f}_\nu^{\text{v}} \right\} \tag{7.38}$$

where $\boldsymbol{H}_\nu^{\text{v}} \in \mathbb{R}^{n_{\text{av}} \times 2}$ and $\boldsymbol{f}_\nu^{\text{v}} \in \mathbb{R}^{n_{\text{av}} \times 1}$. n_{av} is the number of halfspaces used in the approximation. As the voltage bounds do not have to be strictly satisfied in a practical application, the soft constraint set

$$\mathcal{V}_\nu^{\text{s}} = \left\{ \boldsymbol{v}_\nu \in \mathbb{R}^2, \boldsymbol{e}_\nu \in \mathbb{R}^{n_{\text{av}}} | \boldsymbol{H}^{\text{v}} \boldsymbol{v}_\nu \leq \boldsymbol{f}_\nu^{\text{v}} + \boldsymbol{e}_\nu, \boldsymbol{e}_\nu \geq \boldsymbol{0} \right\}. \tag{7.39}$$

is utilized. In combination with the penalization of the soft constraint variables $\boldsymbol{e}_\nu$ in the cost function, a violation of the original constraints (7.38) is possible, but at a high cost, which is why the controller tries to avoid it.

Because the power in-feed or demand of some components cannot be adjusted by the controller, the apparent node power is split into an uncontrollable exogenous and a controllable part, i.e.,

$$\boldsymbol{s}_\nu = \boldsymbol{s}_{\text{ei},\nu} + \boldsymbol{s}_{\text{con},\nu}. \tag{7.40}$$

In the nodes 1 and 3, the power of the consumers cannot be influenced and thus they belong to the fixed part $\boldsymbol{s}_{\text{ei},\nu} = \boldsymbol{s}_{\text{C},\nu}$, $\nu \in \{1, 3\}$. They are modeled through given values which can be obtained from demand profiles of the consumers which are known by the operators. More information is given in Section 7.3.2.

The variable power part $\boldsymbol{s}_{\mathrm{con},\nu} = \boldsymbol{s}_{\mathrm{PV},\nu}$ $\forall \nu \in \{1,3\}$ stems from the in-feed from the solar plants. Here the active power $p_{\mathrm{PV},\nu}$ is assumed to be uncontrollable and is taken from the prediction of the in-feed. The reactive power of the solar plant $q_{\mathrm{PV},\nu}$ can be controlled and thus used to control the voltage in the grid. The apparent power of the inverters is limited $\forall \nu \in \{1,3\}$ by

$$\boldsymbol{s}_{\mathrm{con},\nu,t} \in \tilde{\mathcal{S}}_{\nu,t} = \left\{ \boldsymbol{s}_t = \begin{pmatrix} p_{\mathrm{PV},\nu,t} \\ q_{\mathrm{PV},\nu,t} \end{pmatrix} | p^2_{\mathrm{PV},\nu,t} + q^2_{\mathrm{PV},\nu,t} \leq s^2_{\mathrm{max},\nu} \right\} \tag{7.41}$$

where $s_{\mathrm{max},\nu}$ is the maximum allowed apparent power. It is assumed that the in-feed power from solar panels and the power consumption do not change significantly over the days. In other words, it is assumed that they are periodic with periodicity P, i.e., $p_{\mathrm{PV},\nu,t} = p_{\mathrm{PV},\nu,t+P}$, $\boldsymbol{s}_{\mathrm{ei},\nu,t} = \boldsymbol{s}_{\mathrm{ei},\nu,t+P}$ $\forall t \in \mathbb{N}$, $\nu \in \{1,3\}$.

WDN

The water level h_t in the reservoir of the WDN can be described by the discrete-time integrator

$$h_{t+1} = h_t + \frac{T_\mathrm{s}}{S_\mathrm{r}} \cdot (q_{\mathrm{W},t} - d_{\mathrm{W},t}) \tag{7.42}$$

$\forall t \in \mathbb{N}$ where T_s is the sampling interval, S_r the base area of the water tank, $q_{\mathrm{W},t} \geq 0$ the water flow into the reservoir and $d_{\mathrm{W},t}$ represents the water demand at the time step t. The water flow $q_{\mathrm{W},t}$ can be controlled by the pump. Thereby, active power is drawn from the grid. The relation is linearly approximated by $q_{\mathrm{W},t} = c_\mathrm{p} \cdot p_{2,t}$ where c_p is a constant which can be determined from the pump characteristics provided by the pump manufacturer. The water demand $d_{\mathrm{W},t}$ is an uncontrollable exogenous input. As the water demand does not change significantly from day to day, it can be approximated with measurement data from the WDN operator. Hence, it is assumed to be periodic, i.e. $d_{\mathrm{W},t} = d_{\mathrm{W},t+P}$ for all $t \in \mathbb{N}$. The water reservoir has a limited capacity and is thus subject to the constrains

$$h_{\mathrm{min}} \leq h_t \leq h_{\mathrm{max}} \tag{7.43}$$

for all $t \in \mathbb{N}$ where h_{min} and h_{max} are minimum and maximum reservoir level. Moreover, it is required that

$$q_{\mathrm{W},t} \in \mathcal{Q} = \{q \in \mathbb{R} | q = 0 \vee q_{\mathrm{min}} \leq q \leq q_{\mathrm{max}}\} \tag{7.44}$$

for all $t \in \mathbb{N}$ with $0 < q_{\mathrm{min}} \leq q_{\mathrm{max}}$. This means that the pumps are either switched off, i.e., $q_\mathrm{W} = 0$ or when they are switched on, a the flow cannot be go below the minimum flow q_{min}. Note that the constraint is non-convex.

Similar to before, the power drawn from the pump is limited by

$$\boldsymbol{s}_{\text{con},2} \in \tilde{\mathcal{S}}_2 = \left\{ \boldsymbol{s}_2 = \begin{pmatrix} p_2 \\ q_2 \end{pmatrix} | p_2^2 + q_2^2 \leq s_{\text{max},2}^2 \right\}, \tag{7.45}$$

as the pump is connected to the grid through an inverter. The reactive power is controlled to zero $q_2 = 0$. There is no uncontrollable exogenous power input in node 2, thus $\boldsymbol{s}_{\text{ei},2} = \boldsymbol{0}$ for all $t \in \mathbb{N}$.

Overall Model

The power grid is now recast in the more general representation (7.1) - (7.3). The inputs are the controllable apparent power, i.e., $\boldsymbol{u}_\nu = \boldsymbol{s}_{\text{con},\nu}\ \forall \nu \in \mathbb{Z}_{1:3}$.

The reservoir dynamics (7.42) can be written as

$$\boldsymbol{x}_{2,t+1} = \boldsymbol{A}_{22,t}\boldsymbol{x}_{2,t} + \boldsymbol{B}_{22,t}\boldsymbol{u}_{2,t} + \boldsymbol{d}_{2,t}, \tag{7.46}$$

$\forall t \in \mathbb{N}$ with the state $\boldsymbol{x}_{2,t} = h_t$, known exogenous input $\boldsymbol{d}_{2,t} = -\frac{T_\text{s}}{S_\text{r}} d_{\text{W},t}$, system matrix $\boldsymbol{A}_{22} = 1$ and input matrix $\boldsymbol{B}_{22} = \frac{T_\text{s}}{S_\text{r}} \cdot c_\text{B}[1\ 0]$ for all $t \in \mathbb{N}$. Subsystems $\mathscr{S}_1$ and $\mathscr{S}_3$ do not contain storages. The structure is decoupled, i.e., $\mathcal{N}_\nu = \emptyset$ for all $\nu \in \mathbb{Z}_{1:3}$.

The water level is subject to the constraints $\boldsymbol{x}_{2,t} \in \mathcal{X}_{2,t}\ \forall t \in \mathbb{N}$ where $\boldsymbol{x}_{2,t} \in \mathcal{X}_{2,t} = \{\boldsymbol{x}_2 | \boldsymbol{x}_{\text{min},2} \leq \boldsymbol{x}_2 \leq \boldsymbol{x}_{\text{max},2}\}$ for all $t \in \mathbb{N}$ with $\boldsymbol{x}_{\text{min},2} = h_{\text{min}}$ and $\boldsymbol{x}_{\text{max},2} = h_{\text{max}}$.

The power constraints $\tilde{\mathcal{S}}_{\nu,t}$ in (7.41) are approximated with the polytopic set $\mathcal{S}_{\nu,t}$ such that $\mathcal{S}_{\nu,t} \subseteq \tilde{\mathcal{S}}_{\nu,t}$ and recast as $\boldsymbol{u}_{\nu,t} \in \mathcal{U}_{\nu,t}$ for $\nu \in \{1,3\}$. Note the sets are periodically time-varying as they depend on the infeed of $p_{\text{PV},\nu,t}$. For subsystem $\mathscr{S}_2$, the constraint $\boldsymbol{u}_{2,t} \in \mathcal{U}_2$ is obtained from the intersection of the polytopic inner-approximation $\mathcal{S}_2 \subseteq \tilde{\mathcal{S}}_2$ and the set $c_\text{p}^{-1}\mathcal{Q}$ in the flow constraint (7.44).

A direct relationship between apparent powers and the voltages is obtained by substituting (7.35a) in (7.35b), i.e., $\boldsymbol{v} = \boldsymbol{D}_{\text{PF}}\boldsymbol{s} + \boldsymbol{d}_{\text{PF}}$ where $\boldsymbol{D}_{\text{PF}} = -\boldsymbol{C}_{\text{PF}}\boldsymbol{A}_{\text{PF}}^{-1}\boldsymbol{B}_{\text{PF}}$ and $\boldsymbol{d}_{\text{PF}} = -\boldsymbol{C}_{\text{PF}}\boldsymbol{A}_{\text{PF}}^{-1}\boldsymbol{c}_{\text{PF}} + \boldsymbol{v}_0$. Recall that $\boldsymbol{A}_{\text{PF}}$ has full rank. Introducing the global voltage set $\mathcal{V}^\text{s} = \prod_{\nu=1}^{M} \mathcal{V}_\nu^\text{s}$ and the global soft constraint variable $\boldsymbol{e} = \text{col}_{\nu\in\mathbb{Z}_{1:M}}(\boldsymbol{e}_\nu)$, the global voltage soft constraint is of the form

$$\left(\boldsymbol{D}_{\text{PF}}\boldsymbol{u} + \tilde{\boldsymbol{d}}_{\text{PF}}, \boldsymbol{e}\right) \in \mathcal{V}^\text{s}, \quad \boldsymbol{e} \geq \boldsymbol{0} \tag{7.47}$$

with $\tilde{\boldsymbol{d}}_{\text{PF}} = \boldsymbol{d}_{\text{PF}} + \boldsymbol{D}_{\text{PF}}\boldsymbol{s}_{\text{ei}}$ which can again be recast as $(\boldsymbol{x}_t, \boldsymbol{u}_t) \in \mathcal{Z}_t^\text{C}$. Note that the coupling between the subsystem in (7.47) arises due to the power flow equations (7.35). The number of coupling constraints is $n_\text{c} = 1$ and the structure is $\mathcal{N}^{\text{CA}} = \emptyset$ and $\mathcal{N}^{\text{CU}} = \mathbb{Z}_{1:3}$.

The soft constraint variables $\boldsymbol{e}$ can be integrated into the Algorithm 4 with slight modifications. Copies of them are included in every subsystem. The solution of subsystem $\mathscr{S}_{i^*}$ is broadcast to all other subsystems in step 5 and is accepted there. This communication step is, however, not required as each subsystem can reconstruct the solution of subsystems $\mathscr{S}_{i^*}$ since all powers $\boldsymbol{s}_{\text{con},\nu}$ are available and constraint (7.47) is known in every subsystem. This reconstruction step can be executed between step 6 and 7 by minimizing the cost $\mathbf{1}^{\mathrm{T}}\boldsymbol{e}_r$ subject to $\left(\boldsymbol{D}_{\mathrm{PF}}(\hat{\boldsymbol{u}}_r^{k+1} + \boldsymbol{s}_{\mathrm{ei},r}) + \boldsymbol{d}_{\mathrm{PF}}, \boldsymbol{e}_r\right) \in \mathcal{V}^{\mathrm{s}}$ for all $r \in \mathbb{Z}_{0:N-1}$. Note that for the considered set-up this results in solving N computationally cheap LPs.

The stage cost is composed of three parts. The first part is

$$l^{\mathrm{x}}(\boldsymbol{x}) = (\boldsymbol{x} - \boldsymbol{x}_{\mathrm{ref}})^{\mathrm{T}} \boldsymbol{Q}_{\mathrm{x}} (\boldsymbol{x} - \boldsymbol{x}_{\mathrm{ref}})$$

where $\boldsymbol{x}_{\mathrm{ref}}$ is a reference value for the state. The second part is $l^{\mathrm{u}}(\boldsymbol{u}) = \boldsymbol{u}^{\mathrm{T}}\boldsymbol{Q}_{\mathrm{u}}\boldsymbol{u}$. The third part is $l^{\mathrm{e}}(\boldsymbol{e}) = \boldsymbol{e}^{\mathrm{T}}\boldsymbol{Q}_{\mathrm{e}}\boldsymbol{e}$. $\boldsymbol{Q}_{\mathrm{x}}, \boldsymbol{Q}_{\mathrm{s}}$ and $\boldsymbol{Q}_{\mathrm{e}}$ are positive semidefinite weighting matrices. The total stage cost can be recast as $l(\boldsymbol{x}, \boldsymbol{u}, \boldsymbol{e}) = l^{\mathrm{x}}(\boldsymbol{x}) + l^{\mathrm{u}}(\boldsymbol{u}) + l^{\mathrm{e}}(\boldsymbol{e})$. Optimization problem (7.5) can be recast as a MIQP.

The set-up is tested with two controllers which are located in subsystem 1 and 2. The controller of subsystem 1 optimizes over the inputs $\boldsymbol{u}_1$ and $\boldsymbol{u}_2$, i.e., $\mathcal{N}_1^{\mathrm{U}} = \{1, 2\}$, and the controller of subsystem 1 optimizes over the inputs $\boldsymbol{u}_2$ and $\boldsymbol{u}_3$, i.e., $\mathcal{N}_2^{\mathrm{U}} = \{2, 3\}$. Thus, the set of states that are directly affected are $\mathcal{N}_1^{\mathrm{X}} = \{2\}$ and $\mathcal{N}_2^{\mathrm{X}} = \{2\}$.

As pointed out in Remark 7.1.3, in case of a failure in one of one of the controllers, a part of the system can still be controlled. For the considered set-up, subsystem 1 and 2 could be still be operated in case of a failure of the controller in subsystem 2. Moreover, subsystem 2 and 3 could be operated in case of a failure of the controller in subsystem 1.

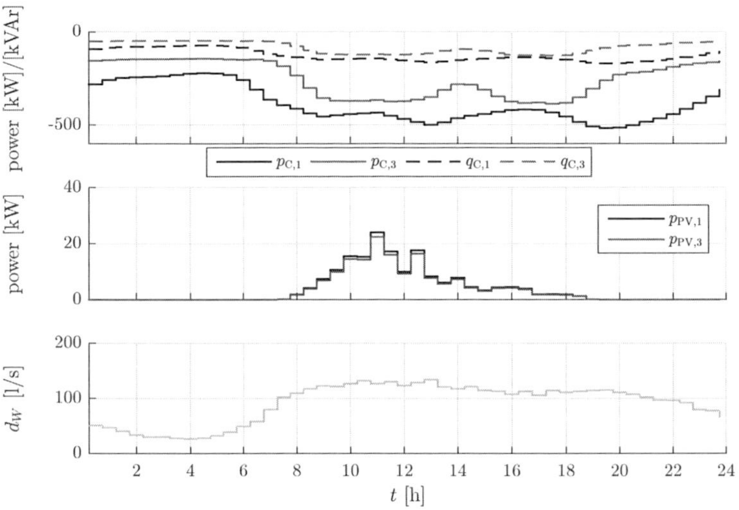

Figure 7.10: Consumer power demand, in-feed from solar plants and water consumption over one day

7.3.2 Simulation Example

Simulation Set-up

Next, the controller is tested in simulation. The grid data can be found in Table 7.3. The minimum and maximum voltage are chosen $\pm 2\,\%$ of the nominal voltage.

In the top figure of Figure 7.10, the active and reactive consumer power $p_{\mathrm{C},\nu}$ and $q_{\mathrm{C},\nu}$ are shown. In the middle part of Figure 7.10 the active power in-feed $p_{\mathrm{PV},\nu}$ for $\nu \in \{1,3\}$ from the solar plants are depicted. The in-feeds are almost the same as the solar plants are located close to each other but have a slightly different peak power. The data is taken from a winter day where there is a high power demand and low infeed from the renewable energy sources so that the lower voltage limitation is violated in the uncontrolled case over the day. In the bottom figure, the water demand d_{W} is shown. The data is taken from measurements. The in-feed and consumer powers as well as the water demand are assumed to be periodic, i.e., are the same for every day. The weighting matrices are $\boldsymbol{Q}_{\mathrm{x}} = 0.01, \boldsymbol{Q}_{\mathrm{u}} = \mathrm{diag}_{\nu \in \mathbb{Z}_{1:M}} \left(\boldsymbol{Q}_{\mathrm{u}\nu}\right), \boldsymbol{Q}_{\mathrm{e}} = 10^6 \cdot \boldsymbol{I}$ with $\boldsymbol{Q}_{\mathrm{s1}} = \boldsymbol{Q}_{\mathrm{s2}} = 1 \cdot \boldsymbol{I}$, $\boldsymbol{Q}_{\mathrm{s3}} = 10 \cdot \boldsymbol{I}$. Note that the cost for reactive power in node 3 is higher so that the input of node 1 is preferred. An in-feed of node 3 increases the losses in the branches because of the larger distance to the slack bus node. The matrix $\boldsymbol{Q}_\mathrm{e}$ weights the penalization of the violation of the voltage constraints. The weight is chosen relatively high so that a violation is avoided. For $t = 0$ initially feasible sequences $\hat{\boldsymbol{X}}_0^1$ and $\hat{\boldsymbol{U}}_0^1$ have to be found. This is done by setting $p_{2,r} = \frac{1}{c_\mathrm{B}} d_{\mathrm{W},r}$, $q_{1,r} = q_{3,r} = 0$ and $\boldsymbol{x}_{r|r} = \boldsymbol{x}_0 \; \forall r \in \mathbb{Z}_{0:N-1}$. Then, each subsystem determines the soft constraint variables. The approach generates feasible solutions as all constraints in (7.5) are satisfied.

For the computation of the voltages in simulation, the set of nonlinear equations (7.34) is solved using the FBS method. In the controller the approximated model (7.35) is applied.

Table 7.3: Parameters of the investigated MV power grid and the connected WDN

description	name	value
line resistance	R'	$0.264\,\Omega$
line inductance	X'_L	$0.189\,\mathrm{mH}$
line capacitance	X'_C	$0.27\,\mu\mathrm{F}$
line length slack bus to node 1	l_{01}	$8\,\mathrm{km}$
line length node 1 to 2	l_{12}	$4\,\mathrm{km}$
line length node 2 to 3	l_{23}	$4\,\mathrm{km}$
base area	S_r	$833.3\,\mathrm{m}^2$
max. tank level	h_max	$5.65\,\mathrm{m}$
min. tank level	h_min	$3\,\mathrm{m}$
power factor	c_p	$-519.73\,\frac{1}{\mathrm{Ws}}$
min. flow	q_min	$10\,\frac{1}{\mathrm{s}}$
max. flow	q_max	$210\,\frac{1}{\mathrm{s}}$
sampling interval	T_s	$30\,\mathrm{min}$
periodicity	P	48
prediction horizon	N	24
max. iterations	k_max	10
threshold	η_i	$-1 \cdot 10^{-3}$
max. apparent power node 1	$s_{\mathrm{max},1}$	$250\,\mathrm{kVA}$
max. apparent power node 2	$s_{\mathrm{max},2}$	$118.5\,\mathrm{kVA}$
max. apparent power node 3	$s_{\mathrm{max},3}$	$233\,\mathrm{kVA}$
max. voltage	v_max	$11.78\,\mathrm{kV}$
min. voltage	v_min	$11.32\,\mathrm{kV}$

Simulation Results

In Figure 7.11 the uncontrolled nodal voltages are shown. Thereby, the in-feed of reactive power from the PV inverters are set to zero, i.e. $q_{\mathrm{PV},1,t} = q_{\mathrm{PV},3,t} = 0$, and the power $p_{2,t} = \frac{1}{c_{\mathrm{B}}} d_{\mathrm{W},t}$ $t \in \mathbb{Z}_{0:47}$. Obviously, a controller is needed to prevent the violation of the lower voltage boundaries in node 2 and 3 from approx. 9 a.m. to 2 p.m. and from approx. 3 p.m. to 8 p.m.

The simulation results with controller are shown in the Figure 7.12. It can be observed that the controller respects all the constraints on the voltage, power and water level. Moreover, the water level converges to the optimal sequence $x_t^{\mathrm{P}} = h_t^{\mathrm{P}}$ for $t \in \mathbb{Z}_{0:23}$ obtained by solving (2.24) offline. One can see that the water reservoir is filled over the day before the voltage magnitude approaches the limitation at 8 a.m. Then, no more water is pumped and the water demand is delivered by emptying the reservoir until approx. 2 p.m. where there is a short dip in the power demand (cf. Figure 7.10) which is exploited to fill the tank again. Afterwards it is emptied again. The rest of the day the water flow q_{W} follows the water demand d_{W}. Parallel to the injection of active power, reactive power is injected in the nodes 1 and 3, whereas node 1 injects more reactive power due to the higher costs in node 3. The in-feed of reactive power prevents the nodal voltages from breaking through their lower boundaries. It is injected at the peak times from 9 a.m. to 2 p.m. and from approx. 3 p.m. to 8 p.m. and increases the voltage in this time. The lower subfigure shows the iterations of the DMPC controller. At beginning the subsystems exchange data to improve the overall cost, but after 15 steps no more iterations are required as no sufficient cost decrease can be achieved. The DMPC is also compared to CMPC with full model information. The closed-loop cost $J = \sum_{t=0}^{T_{\mathrm{sim}}} l_t(\boldsymbol{x}_t, \boldsymbol{u}_t)$ are $J_{\mathrm{CMPC}} = 0.1713\,\mathrm{pu}$ and $J_{\mathrm{DMPC}} = 0.2012\,\mathrm{pu}$ showing the DMPC is performing worse due to the distributed decision making and the incomplete knowledge in the subsystems. Considering the mean computational time of both controllers $t_{\mathrm{CMPC}} = 53.2\,\mathrm{ms}$ and $t_{\mathrm{DMPC}} = 47.7\,\mathrm{ms}$, it can be seen that the CMPC controller needs in average more time than the DMPC.

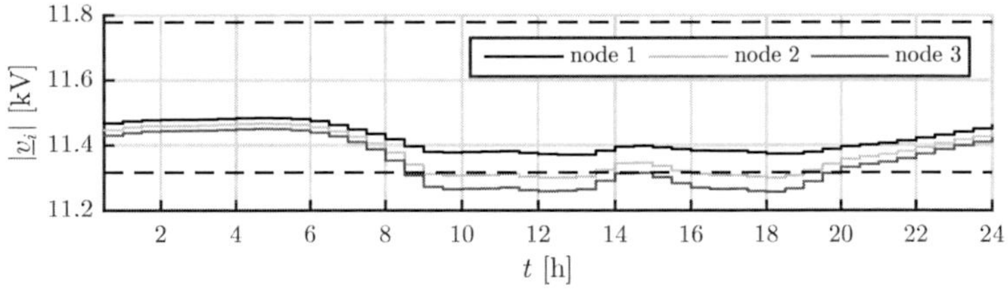

Figure 7.11: Uncontrolled nodal voltages

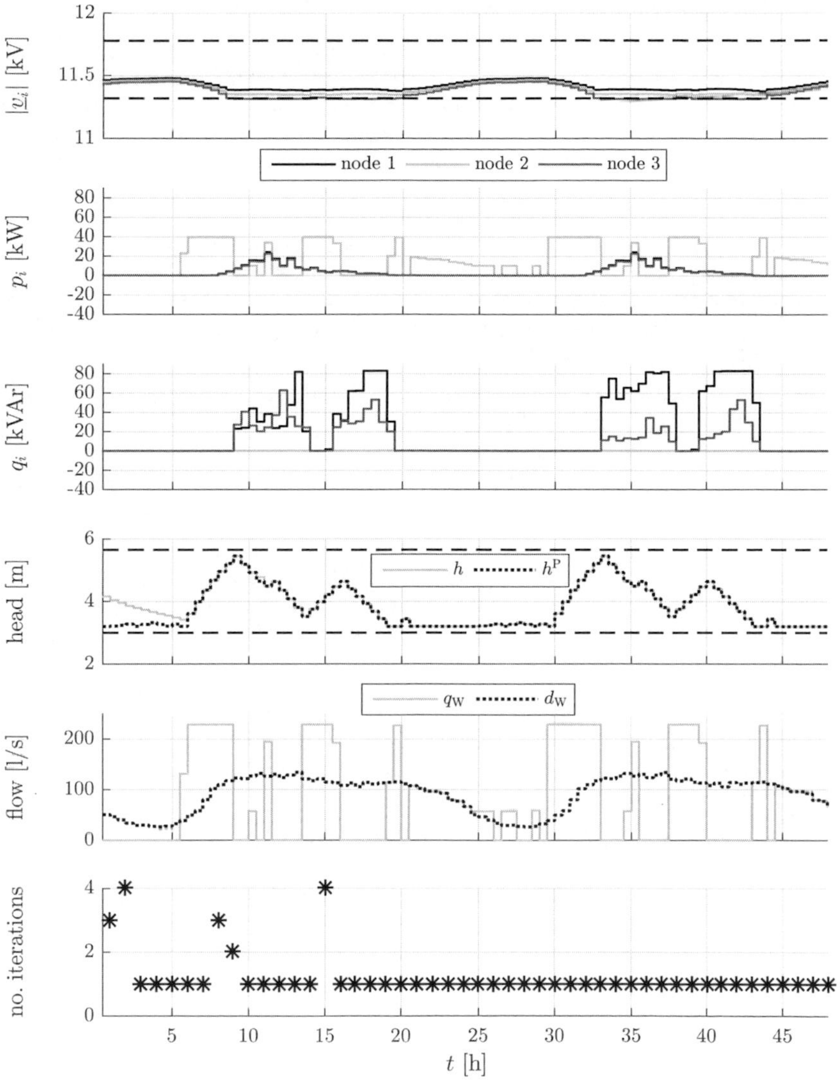

Figure 7.12: Simulation results of DMPC applied to considered MV power grid

7.4 Summary

In this chapter an algorithm for economic DMPC of LPTV subsystems subject to non-convex constraints is developed which is inspired by two systems derived from real-world applications. The algorithm exploits the idea of event-triggered communication for reducing the communication effort. Events are triggered when a controller can contribute significantly to an overall cost decrease. Through sequential updates, it is ensured that the algorithm generates feasible and cost decreasing solutions in every iteration. Moreover, the controller exploits a previously known optimal periodic sequence as terminal constraint for ensuring recursive feasibility and to generate initially feasible input sequences for the next time step after the algorithm terminates. The algorithm can be applied to systems consisting of subsystem which are coupled in the dynamics via inputs and are subject to common constraints.

The algorithm is applied to a control-oriented model of a WDN in Kaiserslautern. The control-oriented model is derived from a detailed nonlinear model and reflects many system properties such as the losses in the system, the pump characteristics and the switching of the pumps. The resulting subsystems are coupled in the dynamics via the inputs and in the constraints. Both coupling types have a cascaded structure.

For the considered simulation scenario, the state and input trajectories converge to the optimal periodic sequence determined offline. The influence of the threshold on the number of iterations and the closed-loop performance is investigated. The performance degradation compared to a centralized implementation is relatively small. If the plant is subject to disturbances, recursive feasibility cannot be ensured due to the absence of inherent robustness properties of the controller. Therefore, a "practical" extension is presented for the WDN application where the hard constraints on the reservoir levels are replaced by soft constraints and the terminal constraint is relaxed to a terminal cost. The extended algorithm is tested for the nonlinear model of the WDN and recursive feasibility is maintained in simulation.

The algorithm is also applied to a model derived from a part of the MV power grid in Kaiserslautern where controllers at several nodes match the infeed from renewable energy sources and a controllable load, a pump station of a WDN which belongs to the same operator as the operator of the power grid, in oder to avoid violation of the desired voltage limitations. The system can be modeled as dynamically decoupled LPTV subsystems subject to decoupled non-convex constraints. Moreover, the subsystems are coupled via a common constraint. The simulation results show that the voltages can be hold in the desired limitations with the controller. It is shown that this is not possible in the uncontrolled case for the considered scenario. The state converges to the optimal periodic sequence. The performance is compared to a centralized implementation.

8 Conclusions and Outlook

8.1 Conclusions

In this thesis MPC strategies for NCSs and distributed control set-ups are studied with a focus on the reduction of the communication effort using event-triggered communication.

In Chapter 4 a method for decentralized OB-MPC with event-triggered communication over the S2C and C2A channels of the network of an NCS is discussed. Discrete-time and sampled-data systems with bounded time-varying transmission delays and sampling intervals are considered in this context.

Chapter 5 deals with a non-iterative DMPC method where the exchange of the state information between the controllers of the subsystems is event-triggered.

Chapter 6 studies a decomposition-based DMPC approach where the cooperation between the controllers of the subsystems for solving a distributed optimization problem is event-triggered.

In Chapter 7 a DMPC approach tailored to two set-ups derived from real-world applications is investigated. The proposed algorithm makes use of event triggers in order to reduce the communication effort.

An overview of the investigated set-ups and control schemes can be found in Table 8.1. While most of the points are self-explanatory or have already been discussed in detail in the previous chapters, some similarities and differences between the approaches should be emphasized in the following.

The approaches of Chapter 4 to 6 rely on robust control methods, which (besides the uncertainty due to the modeling of the couplings in Chapter 4 and 5 and the disturbance in Chapter 6) can be attributed to the event-triggered communication or cooperation. With its help recursive feasibility and stability guarantees can be provided assuming that sufficient conditions are satisfied. The sequential update style of the algorithm in Chapter 7, which is mainly used due to the non-convex constraints and special coupling, avoids a robust design. Recursive feasibility can be provided in this case. Despite missing stability guarantees, the state converged to the optimal periodic sequences in simulation for both set-ups.

The monitored value by the event trigger described in Chapter 7 is the cost which is mainly motivated by the possibility to easily evaluate parts of the cost due to the special coupling topology and type of dynamic coupling. The approaches in Chapter 4 and 5 employ a threshold in the event triggers with relative and absolute part in order to establish attractivity of the origin and recursive feasibility, respectively. Although the considered set-ups are different, the design of controllers is similar. In Chapter 6 a pure absolute threshold is applied.

The simulation results in Chapter 4 and 5 show that for the considered set-ups the communication effort can be reduced significantly compared to their time-triggered equivalents while the performance degradation is relatively small. Similar results are obtained in Chapter 6 when comparing the event-triggered cooperation algorithm to a fully cooperative one. The results in Chapter 7 show that the performance is comparable to the one of a centralized controller.

Table 8.1: Overview of considered plant model, controller set-up and event-trigger properties for different approaches in this thesis.

Chapter	4	5	6	7
Plant Model	discrete-time & sampled-data LTI	discrete-time LTI	discrete-time LTI	discrete-time LPTV
Network-induced effects	yes	-	-	-
Dynamic coupling	state & inputs	state	state	input/-
Constraint coupling	-	-	-	mixed/input
Coupling topology	N2N	N2N	N2N	cascaded/A2A
Constraints	convex	convex	convex	non-convex
Feedback	output	state	state	state
Control architecture	decentralized	distributed	distributed	distributed
Objective function	regulation	regulation	regulation	economic
Controller attitude	non-cooperative	non-cooperative	cooperative	cooperative
Communication style	-	non-iterative	iterative	iterative
Update style	-	parallel	parallel	sequential
Robust MPC	yes	yes	yes	no
Recursive feasibility	yes	yes	yes	yes
Stability guarantees	yes	yes	yes	no
Purpose event trigger	communication S2C and C2A	communication N2N	cooperation N2N	communication N2N/A2A
Exchanged value	output value & input sequence	state value	state sequence	input sequence
Monitored value	output & state	state	state	cost
Threshold	relative & absolute	relative & absolute	absolute	absolute

8.2 Outlook

Consideration of Network-induced Effects in Event-triggered MPC

Except from the investigations in Chapter 4 where bounded time-varying sampling intervals and transmission delays are discussed, it is assumed throughout the thesis that the network is ideal and all components are synchronized. Due to the versatility of imperfections and distributed set-ups, there is a variety of possibilities for the combined consideration of event-triggered control and network-induced imperfections in the future. A few possible research directions are described below.

One option would, for instance, be the consideration of packet dropouts in the S2C channels in the NCS set-up for the OB-MPC scheme presented in Chapter 4. However, the case of packet dropouts in combination with event-triggered output-based control is non-trivial. This observation is also made in [DH17] where an event-triggered non-MPC scheme with packet dropouts in the S2C channel is designed. An analysis of the effects of packet dropouts in the S2C channel on the OB-MPC set-up could be done with the approach in [LS13b]. For a constructive design of event triggers which are aware of packet dropouts, however, more sophisticated event-triggering mechanisms in the sensor unit would be required. One could think of the use of an additional predictor in the sensor unit and an additional information exchange from controller to sensor unit to predict the observers behavior there. Good starting points for future research could be the results in [DH17] and [IFM17].

Many network-induced imperfections in an NCS can also be described by stochastic models, such as Bernoulli or Markov processes. The consideration of stochastic models for variable but bounded transmission delays in combination with linear event-triggered control has shown appealing results, e.g., in [AGL15]. For MPC set-ups with random packet dropouts, the work of [QMFC15] could be a starting point for future investigations.

Considering DMPC with dynamically coupled subsystems, especially, the impact of transmission delays or packet dropouts in the network between the controllers is of interest. The results of [FMP$^+$08, ABB11, LS13a] could be helpful for further research. In addition, compensation approaches proposed for centralized set-ups as in [PP11, VF11] could be extended to distributed schemes.

Another aspect to be taken into account is the assumption of synchronized components in the network. The usage of synchronization protocols leads to increased requirements on the network. It, therefore, makes sense to develop control schemes that are not based on this assumption. For the handling of asynchronous sampling in spatially distributed sensors, the approach of [MC14] could be interesting.

Nonlinear Models

Throughout the thesis, it is dealt with linear time-invariant and periodically time-varying systems. The extension of the methods to systems with nonlinear dynamics is desirable. For the approaches in Chapter 7 this should be straightforward due to the sequential updates in the algorithm. For the approaches relying on robust MPC, the results in [LAC02] could be helpful. In the method presented in Chapter 4, a nonlinear observer that bounds the error between the actual and the observer state would be required which is a challenging task for general nonlinear systems. For special nonlinear systems, such as systems with dominant linear behavior, the design in Chapter 4 could be adapted.

Output-based DMPC with Event-triggered Communication

In many practical applications it is often not possible or desirable to measure the complete state, which is why often only output information is available. As already mentioned in the introductions of Chapters 5 and 6, the output-based case can also handled, e.g., by applying the methods presented in Chapter 4. For the set-up considered in Chapter 7, a straightforward adaption is not possible due to the non-convex constraints and further research is required.

Event-triggered Communication and Stopping Criteria for Distributed Optimization

As pointed out several times within this thesis, decomposition-based schemes require many iterations and hence the communication effort can be demanding. An interesting approach is presented in [MUA14] where events are triggered from an optimization algorithm's point of view. In [RG17] the idea has been applied to the fast gradient method for solving distributed optimization problems arising in DMPC. Here, the development of algorithms suited for DMPC that converge under milder assumptions, such as convex cost functions, would be interesting. Moreover, a bound on the required iterations for achieving a desired suboptimality for such algorithms would be of great interest.

Besides the iterative exchange of optimization variables, the DMPC schemes based on distributed optimization require the exchange of additional messages for checking the stopping criteria of the algorithms. Stopping conditions that can be checked locally or with less communication overhead are appealing.

A Supplementary Material

A.1 Quadratic Matrix Forms

Lemma A.1.1. *Let $\boldsymbol{M} \in \mathbb{R}^{n\times n}$ be $\boldsymbol{M} \succ \boldsymbol{0}$, then it holds for all $\boldsymbol{x} \in \mathbb{R}^n$ that*

$$\|\boldsymbol{x}\|^2 \lambda^{\mathrm{m}}(\boldsymbol{M}) \leq \boldsymbol{x}^{\mathrm{T}}\boldsymbol{M}\boldsymbol{x} \leq \|\boldsymbol{x}\|^2 \lambda^{\mathrm{M}}(\boldsymbol{M}). \tag{A.1}$$

The inequality is also referred to as the Rayleigh-Ritz inequality. Its proof can be found, e.g., in [Mey00, Example 7.5.1].

Lemma A.1.2. *Consider $\boldsymbol{\Theta} \in \mathbb{R}^{m\times n}$, $\eta \in \mathbb{R}_{>0}$ and $\boldsymbol{\Gamma} \in \mathbb{R}^{n\times n}$ with $\boldsymbol{\Gamma} \succ \boldsymbol{0}$, then it holds for all $\boldsymbol{x} \in \mathbb{R}^n$ and $\boldsymbol{y} \in \mathbb{R}^m$ that*

$$\pm 2\boldsymbol{y}^{\mathrm{T}}\boldsymbol{\Theta}\boldsymbol{x} \leq \eta\boldsymbol{x}^{\mathrm{T}}\boldsymbol{\Gamma}\boldsymbol{x} + \frac{1}{\eta}\boldsymbol{y}^{\mathrm{T}}\boldsymbol{\Theta}\boldsymbol{\Gamma}^{-1}\boldsymbol{\Theta}^{\mathrm{T}}\boldsymbol{y}. \tag{A.2}$$

Proof.

$$\begin{aligned}
\pm 2\boldsymbol{y}^{\mathrm{T}}\boldsymbol{\Theta}\boldsymbol{x} &= \pm 2\boldsymbol{y}^{\mathrm{T}}\boldsymbol{\Theta}\boldsymbol{x} - \eta\boldsymbol{x}^{\mathrm{T}}\boldsymbol{\Gamma}\boldsymbol{x} - \frac{1}{\eta}\boldsymbol{y}^{\mathrm{T}}\boldsymbol{\Theta}\boldsymbol{\Gamma}^{-1}\boldsymbol{\Theta}^{\mathrm{T}}\boldsymbol{y} + \eta\boldsymbol{x}^{\mathrm{T}}\boldsymbol{\Gamma}\boldsymbol{x} + \frac{1}{\eta}\boldsymbol{y}^{\mathrm{T}}\boldsymbol{\Theta}\boldsymbol{\Gamma}^{-1}\boldsymbol{\Theta}^{\mathrm{T}}\boldsymbol{y} && \text{(A.3)}\\
&= -\frac{1}{\eta}\left(\eta\boldsymbol{\Gamma}\boldsymbol{x} \mp \boldsymbol{\Theta}^{\mathrm{T}}\boldsymbol{y}\right)^{\mathrm{T}}\boldsymbol{\Gamma}^{-1}\left(\eta\boldsymbol{\Gamma}\boldsymbol{x} \mp \boldsymbol{\Theta}^{\mathrm{T}}\boldsymbol{y}\right) + \eta\boldsymbol{x}^{\mathrm{T}}\boldsymbol{\Gamma}\boldsymbol{x} + \frac{1}{\eta}\boldsymbol{y}^{\mathrm{T}}\boldsymbol{\Theta}\boldsymbol{\Gamma}^{-1}\boldsymbol{\Theta}^{\mathrm{T}}\boldsymbol{y} && \text{(A.4)}\\
&\leq \eta\boldsymbol{x}^{\mathrm{T}}\boldsymbol{\Gamma}\boldsymbol{x} + \frac{1}{\eta}\boldsymbol{y}^{\mathrm{T}}\boldsymbol{\Theta}\boldsymbol{\Gamma}^{-1}\boldsymbol{\Theta}^{\mathrm{T}}\boldsymbol{y}. && \text{(A.5)}
\end{aligned}$$

The inequality holds as the inverse $\boldsymbol{\Gamma}^{-1}$ is a positive definite matrix and thus the first term is negative for all $\boldsymbol{x} \in \mathbb{R}^n$ and $\boldsymbol{y} \in \mathbb{R}^m$. □

A.2 Implementation

The numerical experiments are performed with MATLAB®2014b. The toolbox YALMIP [Löf04] and the solver MOSEK [ApS17] are used for solving the LMI problems. The PnP-toolbox [RBFT13] is utilized for the computation of $\epsilon-$approximations of the minimum RPI sets. All other set computation are carried out with the Multi-Parametric Toolbox 3.0 [HKJM13]. The MATLAB® toolbox of IBM® ILOG® CPLEX® Optimization Studio 12.6 [IBM14] is used to solve the MIQPs and QPQCs.

The computations in Chapter 7 are carried out using an Intel® Core™ i5-660 processor with 3.33 GHz and 8 GB RAM under Microsoft® Windows 7 Professional® .

B Proofs

B.1 Chapter 3

Proof of Lemma 3.4.1

Proof. The proof uses ideas from suboptimal MPC in [RM09, Chapter 6.1.2] and robust MPC in [BHA17].

$V_N(\boldsymbol{x}_{t+1}, \boldsymbol{V}_{t+1})$ can be bounded by

$$V_N(\boldsymbol{x}_{t+1}, \boldsymbol{V}_{t+1}) = \min_{\boldsymbol{z}\in\mathcal{Z}} V^{\mathrm{f}}(\boldsymbol{x}_{t+1} - \boldsymbol{z}) + \nu J_N(\boldsymbol{V}_{t+1}) \tag{B.1}$$

$$\leq \min_{\boldsymbol{z}\in\mathcal{Z}} V^{\mathrm{f}}(\boldsymbol{x}_{t+1} - \boldsymbol{z}) + \nu J_N(\boldsymbol{V}_t) - \nu \boldsymbol{v}_t^{\mathrm{T}} \boldsymbol{M} \boldsymbol{v}_t \tag{B.2}$$

$$\leq V^{\mathrm{f}}\left(\boldsymbol{x}_{t+1} - \boldsymbol{\zeta}_{t+1}\right) + \nu J_N(\boldsymbol{V}_t) - \nu \boldsymbol{v}_t^{\mathrm{T}} \boldsymbol{M} \boldsymbol{v}_t. \tag{B.3}$$

In the first step, $J_N(\boldsymbol{V}_{t+1}) - J_N(\boldsymbol{V}_t) \leq -\boldsymbol{v}_t^{\mathrm{T}} \boldsymbol{M} \boldsymbol{v}_t$ is used. For obtaining the second inequality,

$$\boldsymbol{\zeta}_{t+1} = \boldsymbol{F}\boldsymbol{z}_t^* + \boldsymbol{w}_t \text{ with } \boldsymbol{z}_t^* = \arg\min\nolimits_{\boldsymbol{z}\in\mathcal{Z}} V^{\mathrm{f}}(\boldsymbol{x}_t - \boldsymbol{z}) \tag{B.4}$$

is substituted. Note that $\boldsymbol{\zeta}_{t+1} \in \mathcal{Z}$ since $\mathcal{Z}$ is an RPI set for the dynamics $\boldsymbol{z}_{t+1} = \boldsymbol{F}\boldsymbol{z}_t + \boldsymbol{w}_t$. Defining $\boldsymbol{\xi}_t = \boldsymbol{x}_t - \boldsymbol{z}_t^*$, for the first summand follows

$$V^{\mathrm{f}}\left(\boldsymbol{x}_{t+1} - \boldsymbol{\zeta}_{t+1}\right) = \left(\boldsymbol{F}\boldsymbol{\xi}_t + \boldsymbol{B}\boldsymbol{v}_t\right)^{\mathrm{T}} \boldsymbol{P} \left(\boldsymbol{F}\boldsymbol{\xi}_t + \boldsymbol{B}\boldsymbol{v}_t\right) \tag{B.5}$$

$$\leq \boldsymbol{v}_t^{\mathrm{T}} \boldsymbol{\Upsilon} \boldsymbol{v}_t + \frac{1}{2} \boldsymbol{\xi}_t^{\mathrm{T}} \boldsymbol{Q} \boldsymbol{\xi}_t + \boldsymbol{\xi}_t^{\mathrm{T}} \boldsymbol{F}^{\mathrm{T}} \boldsymbol{P} \boldsymbol{F} \boldsymbol{\xi}_t \tag{B.6}$$

$$\leq \boldsymbol{v}_t^{\mathrm{T}} \boldsymbol{\Upsilon} \boldsymbol{v}_t - \frac{1}{2} \boldsymbol{\xi}_t^{\mathrm{T}} \boldsymbol{Q} \boldsymbol{\xi}_t + \boldsymbol{\xi}_t^{\mathrm{T}} \boldsymbol{P} \boldsymbol{\xi}_t \tag{B.7}$$

with $\boldsymbol{\Upsilon} = \boldsymbol{B}^{\mathrm{T}}\left(\boldsymbol{P} + 2\boldsymbol{P}\boldsymbol{F}\boldsymbol{Q}^{-1}\boldsymbol{F}^{\mathrm{T}}\boldsymbol{P}\right)\boldsymbol{B}$. The first inequality is derived using (A.2). The second inequality holds due to $\boldsymbol{F}^{\mathrm{T}}\boldsymbol{P}\boldsymbol{F} - \boldsymbol{P} \preceq -\boldsymbol{Q}$. It follows that

$$V_N(\boldsymbol{x}_{t+1}, \boldsymbol{V}_{t+1}) - V_N(\boldsymbol{x}_t, \boldsymbol{V}_t) \leq -\frac{1}{2}\boldsymbol{\xi}_t^{\mathrm{T}}\boldsymbol{Q}\boldsymbol{\xi}_t + \boldsymbol{v}_t^{\mathrm{T}}(\boldsymbol{\Upsilon} - \nu\boldsymbol{M})\boldsymbol{v}_t \tag{B.8}$$

$$= -\min_{\boldsymbol{z}\in\mathcal{Z}} \frac{1}{2}(\boldsymbol{x}_t - \boldsymbol{z})^{\mathrm{T}}\boldsymbol{Q}(\boldsymbol{x}_t - \boldsymbol{z}) + \boldsymbol{v}_t^{\mathrm{T}}(\boldsymbol{\Upsilon} - \nu\boldsymbol{M})\boldsymbol{v}_t \tag{B.9}$$

$$\leq -\frac{1}{2}\lambda^{\mathrm{m}}(\boldsymbol{Q})\|\boldsymbol{x}_t\|_{\mathcal{Z}}^2 + \boldsymbol{v}_t^{\mathrm{T}}(\boldsymbol{\Upsilon} - \nu\boldsymbol{M})\boldsymbol{v}_t. \tag{B.10}$$

Choosing ν such that $\boldsymbol{M} \succ \frac{1}{\nu}\boldsymbol{\Upsilon}$ leads to

$$V_N(\boldsymbol{x}_{t+1}, \boldsymbol{V}_{t+1}) - V_N(\boldsymbol{x}_t, \boldsymbol{V}_t) \leq -c_1\|(\boldsymbol{x}_t, \boldsymbol{v}_t)\|_{\mathcal{Z}\times\{\boldsymbol{0}\}}^2 \tag{B.11}$$

$\forall t \in \mathbb{N}$ where $c_1 = \min\left(\frac{1}{2}\lambda^{\mathrm{m}}(\boldsymbol{Q}), \lambda^{\mathrm{m}}\left(\boldsymbol{M} - \frac{1}{\nu}\boldsymbol{\Upsilon}\right)\right) > 0$. This implies that V_N is a non-increasing function. As it is bounded below by zero it follows that $\lim_{t\to\infty}\|\boldsymbol{x}_t\|_{\mathcal{Z}} = 0$ as well as $\lim_{t\to\infty}\|\boldsymbol{v}_t\| = 0$. □

Proof of Theorem 3.4.2

Proof. The proof applies ideas from suboptimal MPC in [RM09, Chapter 6.1.2]. It is shown that $V_N(\boldsymbol{x}, \boldsymbol{V})$ is a Lyapunov function for (3.6).

Let $s = \max_{\boldsymbol{V}\in\mathcal{D}_N}\|\boldsymbol{V}\|$. Note that $\mathcal{D}_N$ is compact since $\mathcal{X}$ and $\mathcal{U}$ are compact and thus the constant s exists due to the Weierstrass theorem.

Hence, for $\boldsymbol{x} \in \mathcal{X}_N \setminus \mathcal{B}_\epsilon(\boldsymbol{0})$, $\|\boldsymbol{V}\| \leq \frac{s}{\epsilon}\|\boldsymbol{x}\|_{\mathcal{Z}}$ holds.

Together with (3.10) it follows that

$$\|\boldsymbol{V}_t\| \leq \bar{d}\,\|\boldsymbol{x}_t\|_{\mathcal{Z}} \text{ with } \bar{d} = \max(\frac{s}{\epsilon}, d) \quad \boldsymbol{x}_t \in \mathcal{X}_N. \tag{B.12}$$

Thus,

$$\|(\boldsymbol{x}_t, \boldsymbol{V}_t)\|_{\mathcal{Z}\times\{\boldsymbol{0}\}}^2 = \|\boldsymbol{x}_t\|_{\mathcal{Z}}^2 + \|\boldsymbol{V}_t\|^2 \leq (1 + \bar{d}^2)\|\boldsymbol{x}_t\|_{\mathcal{Z}}^2. \tag{B.13}$$

Together with (3.9) this leads to

$$V_N(\boldsymbol{x}_{t+1}, \boldsymbol{V}_{t+1}) - V_N(\boldsymbol{x}_t, \boldsymbol{V}_t) \leq -c_4\|(\boldsymbol{x}_t, \boldsymbol{V}_t)\|_{\mathcal{Z}\times\{\boldsymbol{0}\}}^2 \tag{B.14}$$

for all $t \in \mathbb{N}$ with $c_4 = c_1/(1 + \bar{d}^2)$.

The function $V_N(\boldsymbol{x}, \boldsymbol{V})$ can be lower and upper bounded by

$$c_2\|(\boldsymbol{x}, \boldsymbol{V})\|_{\mathcal{Z}\times\{\boldsymbol{0}\}}^2 \leq V_N(\boldsymbol{x}, \boldsymbol{V}) \leq c_3\|(\boldsymbol{x}, \boldsymbol{V})\|_{\mathcal{Z}\times\{\boldsymbol{0}\}}^2 \tag{B.15}$$

where $c_2 = \min(\lambda^{\mathrm{m}}(\boldsymbol{P}), \lambda^{\mathrm{m}}(\boldsymbol{M})) > 0$ and $c_3 = \max(\lambda^{\mathrm{M}}(\boldsymbol{P}), \lambda^{\mathrm{M}}(\boldsymbol{M})) > 0$. Thus, according to Theorem 2.2.11 there exist $\iota > 0$ and $0 < \gamma < 1$ such that

$$\|(\boldsymbol{x}_t, \boldsymbol{V}_t)\|_{\mathcal{Z}\times\{\mathbf{0}\}} \leq \iota\gamma^t \|(\boldsymbol{x}_0, \boldsymbol{V}_0)\|_{\mathcal{Z}\times\{\mathbf{0}\}}. \tag{B.16}$$

Moreover, it follows that

$$\|\boldsymbol{x}_t\|_{\mathcal{Z}} \leq \iota\sqrt{1+\bar{d}^2}\gamma^t \|\boldsymbol{x}_0\|_{\mathcal{Z}}. \tag{B.17}$$

This completes the proof. □

B.2 Chapter 4

Proof of Theorem 4.3.1

Proof. The error $\boldsymbol{e}_{i,t}$ is bounded by the event trigger (4.3), i.e., $\boldsymbol{e}_{i,t} \in \mathcal{E}_i^{\mathrm{a}}$ for all $t \in \mathbb{N}$. Assume that $\boldsymbol{x}_{i,t} \in \mathcal{X}_i$ and $\boldsymbol{u}_{i,t} \in \mathcal{U}_i$ for all $t \in \mathbb{N}$ and $i \in \mathbb{Z}_{1:M}$ (it shown below that this assumption holds). It follows that $\boldsymbol{w}_{i,t}^{\mathrm{o}} \in \mathcal{W}_i^{\mathrm{o}}$ is bounded $\forall t \in \mathbb{N}$. By construction of the set Ω_i as an RPI set for the error dynamics (4.6) and together with the assumption that $\tilde{\boldsymbol{x}}_{i,0} \in \Omega_i$, it follows that $\tilde{\boldsymbol{x}}_{i,t} \in \Omega_i$ $\forall t \in \mathbb{N}$, $i \in \mathbb{Z}_{1:M}$. This implies that the disturbance $\boldsymbol{w}_{i,t} \in \mathcal{W}_i$ for all $t \in \mathbb{N}$, $i \in \mathbb{Z}_{1:M}$. It follows from Lemma 2.3.3 that $\mathcal{X}_{N,i}$ is an RPI set for (4.7). Thus, $\hat{\boldsymbol{x}}_0 \in \mathcal{X}_N$ leads to $\hat{\boldsymbol{x}}_t \in \mathcal{X}_N$ and $\boldsymbol{u}_t \in \mathcal{U}$ $\forall t \in \mathbb{N}$. Due to the construction of the sets $\bar{\mathcal{X}}^l$ it follows that $\mathcal{X}_N \subseteq \mathcal{X} \ominus \Omega$ and thus $\boldsymbol{x}_t = \hat{\boldsymbol{x}}_t + \tilde{\boldsymbol{x}}_t \in (\mathcal{X} \ominus \Omega) \oplus \Omega \subseteq \mathcal{X}$ for all $t \in \mathbb{N}$. This completes the proof. □

Proof of Lemma 4.3.3

Proof. (a) It is first shown that Assumption 4.2.3 implies that $\mathcal{X}^{\mathrm{f}} \subseteq \mathcal{X}_K$. Due to the assumption $(\boldsymbol{A}_{\mathrm{D}} + \boldsymbol{B}_{\mathrm{D}}\boldsymbol{K})\mathcal{X}^{\mathrm{f}} \subseteq \mathcal{X}^{\mathrm{f}}$ holds. So for any $\boldsymbol{x} \in \mathcal{X}^{\mathrm{f}}$ this means that $(\boldsymbol{A}_{\mathrm{D}} + \boldsymbol{B}_{\mathrm{D}}\boldsymbol{K})^l \boldsymbol{x} \in \mathcal{X}^{\mathrm{f}}$ for $l \in \mathbb{Z}_{0:N}$ and thus $(\boldsymbol{A}_{\mathrm{D}} + \boldsymbol{B}_{\mathrm{D}}\boldsymbol{K})^l \boldsymbol{x} \in \bar{\mathcal{X}}^l$ as well as
$\boldsymbol{K}(\boldsymbol{A}_{\mathrm{D}} + \boldsymbol{B}_{\mathrm{D}}\boldsymbol{K})^l \boldsymbol{x} \in \bar{\mathcal{U}}^l$ which implies that $\boldsymbol{x} \in \mathcal{X}_K$.

As by assumption $\mathcal{X}^{\mathrm{f}}$ is non-empty this implies that $\mathcal{X}_K$ is non-empty. Since Ω is also non-empty, Φ_K is non-empty.

(b) It is first shown that $\mathcal{X}_K$ is equivalent to the set $\mathcal{X}_0 = \{\boldsymbol{x} \in \mathbb{R}^n | J_N^*(\boldsymbol{x}) = 0\}$. Assume that $\boldsymbol{x} \in \mathcal{X}_N$. From the definition of the set follows that $J_N(\bar{\boldsymbol{V}}) = 0$. From optimality and the fact that J_N is bounded below by zero, it follows that $J_N^*(\boldsymbol{x}) = 0$ and thus that $\boldsymbol{x} \in \mathcal{X}_0$. Now assume that $\boldsymbol{x} \in \mathcal{X}_0$. $J_N^*(\boldsymbol{x}) = 0$ implies that $\boldsymbol{V}^* = 0$ as $\boldsymbol{M} \succ \mathbf{0}$. This implies that $\boldsymbol{x} \in \mathcal{X}_K$.

Now assume $\hat{\boldsymbol{x}}_t \in \mathcal{X}_K$. It follows that $J_N^*(\hat{\boldsymbol{x}}_t) = 0$. From (4.21) it follows that $J_N^*(\hat{\boldsymbol{x}}_{t+1}) = 0$ and thus $\hat{\boldsymbol{x}}_{t+1} \in \mathcal{X}_K$, i.e., $\mathcal{X}_K$ is RPI for (4.17) and (4.19).

(c) As Ω is RPI for system (4.6) and $\mathcal{X}_K$ is RPI for system (4.17), it follows that Φ_K is RPI for the dynamics (4.18) and (4.19). This completes the proof. □

Proof of Lemma 4.3.5

Proof. Multiplying (4.22) from right and left with $\left(\boldsymbol{\phi}_t^{\mathrm{T}} \quad \boldsymbol{e}_t^{\mathrm{T}}\right)^{\mathrm{T}}$ and its transpose, respectively, one obtains

$$\boldsymbol{\phi}_t^{\mathrm{T}}\left(\check{\boldsymbol{P}}-\check{\boldsymbol{A}}_K^{\mathrm{T}}\check{\boldsymbol{P}}\check{\boldsymbol{A}}_K-\delta\check{\boldsymbol{Q}}-\check{\boldsymbol{S}}^{\mathrm{s}}\right)\boldsymbol{\phi}_t-2\boldsymbol{e}_t^{\mathrm{T}}\check{\boldsymbol{A}}_K^{\mathrm{T}}\check{\boldsymbol{P}}\check{\boldsymbol{L}}\boldsymbol{\phi}_t+\boldsymbol{e}_t^{\mathrm{T}}\left(\boldsymbol{S}^{\mathrm{s}}-\check{\boldsymbol{L}}^{\mathrm{T}}\check{\boldsymbol{P}}\check{\boldsymbol{L}}-\check{\boldsymbol{N}}\right)\boldsymbol{e}_t \geq 0. \tag{B.18}$$

The fact that (4.20) holds, implies that

$$\left(\check{\boldsymbol{A}}_K\boldsymbol{\phi}_t+\check{\boldsymbol{L}}\boldsymbol{e}_t\right)^{\mathrm{T}}\check{\boldsymbol{P}}\left(\check{\boldsymbol{A}}_K\boldsymbol{\phi}_t+\check{\boldsymbol{L}}\boldsymbol{e}_t\right)-\boldsymbol{\phi}_t^{\mathrm{T}}\check{\boldsymbol{P}}\boldsymbol{\phi} \leq -\delta\boldsymbol{\phi}_t^{\mathrm{T}}\check{\boldsymbol{Q}}\boldsymbol{\phi}_t-\boldsymbol{e}_t^{\mathrm{T}}\check{\boldsymbol{N}}\boldsymbol{e}_t. \tag{B.19}$$

By assumption $\boldsymbol{\phi}_0 \in \Phi_K$ and the fact that Φ_K is RPI for (4.18) (Lemma 4.3.3 (c)), it follows that $\boldsymbol{u}_t = \boldsymbol{K}\hat{\boldsymbol{x}}_t\ \forall t \in \mathbb{N}$. Thus, it follows that the dynamic evolves according to $\boldsymbol{\phi}_{t+1} = \check{\boldsymbol{A}}_K\boldsymbol{\phi}_t + \check{\boldsymbol{L}}\boldsymbol{e}_t$ and (4.23) holds. This completes the proof. □

Proof of Lemma 4.3.6

Proof. It is shown that σ^{s}, δ and $\boldsymbol{S}^{\mathrm{s}}$ in the LMI can always be selected such that (4.23) holds. As the matrix $\check{\boldsymbol{A}}_K$ is stable and there exists a matrix $\check{\boldsymbol{P}} \succ \boldsymbol{0}$ for a given $\check{\boldsymbol{Q}} \succ \boldsymbol{0}$ such that $\check{\boldsymbol{A}}_K^{\mathrm{T}}\check{\boldsymbol{P}}\check{\boldsymbol{A}}_K - \check{\boldsymbol{P}} = -\check{\boldsymbol{Q}}$. If $\boldsymbol{\phi}_0 \in \Phi_K$ then

$$\Delta V^{\mathrm{f}} = V^{\mathrm{f}}\left(\boldsymbol{\phi}_{t+1}\right) - V^{\mathrm{f}}\left(\boldsymbol{\phi}_t\right) \tag{B.20}$$

$$= -\boldsymbol{\phi}_t^{\mathrm{T}}\check{\boldsymbol{Q}}\boldsymbol{\phi}_t + 2\boldsymbol{e}_t^{\mathrm{T}}\check{\boldsymbol{L}}^{\mathrm{T}}\check{\boldsymbol{P}}\check{\boldsymbol{A}}_K\boldsymbol{\phi}_t + \boldsymbol{e}_t^{\mathrm{T}}\check{\boldsymbol{L}}^{\mathrm{T}}\check{\boldsymbol{P}}\check{\boldsymbol{L}}\boldsymbol{e}_t \tag{B.21}$$

$\forall t \in \mathbb{N}$. Eliminating the cross-terms by applying (A.2) with $\eta = \lambda > 0$, one obtains

$$\Delta V^{\mathrm{f}} \leq (\lambda-1)\boldsymbol{\phi}_t^{\mathrm{T}}\check{\boldsymbol{Q}}\boldsymbol{\phi}_t + \boldsymbol{e}_t^{\mathrm{T}}\left(\Xi - \check{\boldsymbol{N}}\right)\boldsymbol{e}_t \tag{B.22}$$

with $\Xi = \check{\boldsymbol{N}} + \check{\boldsymbol{L}}^{\mathrm{T}}\left(\check{\boldsymbol{P}} + \lambda^{-1}\check{\boldsymbol{P}}\check{\boldsymbol{A}}_K\check{\boldsymbol{Q}}^{-1}\check{\boldsymbol{A}}_K^{\mathrm{T}}\check{\boldsymbol{P}}\right)\check{\boldsymbol{L}}$ for an arbitrary $\check{\boldsymbol{N}} \succ \boldsymbol{0}$. Adding the term $\left(\sigma^{\mathrm{s}}\boldsymbol{\phi}_t^{\mathrm{T}}\check{\boldsymbol{S}}^{\mathrm{s}}\boldsymbol{\phi}_t - \boldsymbol{e}_t^{\mathrm{T}}\boldsymbol{S}^{\mathrm{s}}\boldsymbol{e}_t\right) \geq 0$, leads to

$$\Delta V^{\mathrm{f}} \leq \boldsymbol{\phi}_t^{\mathrm{T}}\left(\sigma^{\mathrm{s}}\check{\boldsymbol{C}}^{\mathrm{T}}\boldsymbol{S}^{\mathrm{s}}\check{\boldsymbol{C}} - (\lambda-1)\check{\boldsymbol{Q}}\right)\boldsymbol{\phi}_t + \boldsymbol{e}_t^{\mathrm{T}}\left(\Xi - \boldsymbol{S}^{\mathrm{s}} - \check{\boldsymbol{N}}\right)\boldsymbol{e}_t. \tag{B.23}$$

Choosing $\boldsymbol{S}^{\mathrm{s}} = \lambda^{\mathrm{M}}(\Xi)\cdot\boldsymbol{I}$, $0 \leq \sigma^{\mathrm{s}} \leq \bar{\sigma}^{\mathrm{s}}$ where $\bar{\sigma}^{\mathrm{s}} = \frac{\frac{1}{2}(1-\lambda-\delta)\lambda^{\mathrm{m}}\left(\check{\boldsymbol{Q}}\right)}{\lambda^{\mathrm{M}}\left(\check{\boldsymbol{C}}^{\mathrm{T}}\boldsymbol{S}^{\mathrm{s}}\check{\boldsymbol{C}}\right)} > 0$, $0 < \delta < 1$ and $1 - \lambda - \delta > 0$, implies that (4.23) holds. This completes the proof. □

Proof of Theorem 4.3.7

Proof. First, it is proven that the function V_N^* is a Lyapunov function for the closed-loop system (4.18) and (4.19). Therefore, the proof is split into three parts. In the first part it is shown that

$$V_N^*\left(\boldsymbol{\phi}_{t+1}\right) - V_N^*\left(\boldsymbol{\phi}_t\right) \leq -c_3 \left\|\boldsymbol{\phi}_t\right\|^2 \quad \forall t \in \mathbb{N} \tag{B.24}$$

with $c_3 > 0$. Computing the difference $\Delta V_N^* = V_N^*\left(\boldsymbol{\phi}_{t+1}\right) - V_N^*\left(\boldsymbol{\phi}_t\right)$ yields

$$\Delta V_N^* \leq -\delta \boldsymbol{\phi}_t^{\mathrm{T}} \check{\boldsymbol{Q}} \boldsymbol{\phi}_t + 2\boldsymbol{v}_t^{\mathrm{T}} \check{\boldsymbol{B}}^{\mathrm{T}} \check{\boldsymbol{P}} \check{\boldsymbol{A}}_K \boldsymbol{\phi}_t + 2\boldsymbol{e}_t^{\mathrm{T}} \check{\boldsymbol{L}}^{\mathrm{T}} \check{\boldsymbol{P}} \check{\boldsymbol{B}} \boldsymbol{v}_t \tag{B.25}$$

$$+ \boldsymbol{v}_t^{\mathrm{T}} \check{\boldsymbol{B}}^{\mathrm{T}} \check{\boldsymbol{P}} \check{\boldsymbol{B}} \boldsymbol{v}_t - \boldsymbol{e}_t^{\mathrm{T}} \check{\boldsymbol{N}} \boldsymbol{e}_t - \nu \boldsymbol{v}_t^{\mathrm{T}} \boldsymbol{M} \boldsymbol{v}_t \tag{B.26}$$

$$\leq -\frac{\delta}{2} \boldsymbol{\phi}_t^{\mathrm{T}} \check{\boldsymbol{Q}} \boldsymbol{\phi}_t - \boldsymbol{v}_t^{\mathrm{T}} \left(\boldsymbol{\Upsilon} - \nu \boldsymbol{M}\right) \boldsymbol{v}_t \tag{B.27}$$

with $\boldsymbol{\Upsilon} = \check{\boldsymbol{B}}^{\mathrm{T}} \Big(\check{\boldsymbol{P}} + \check{\boldsymbol{P}} \Big(\frac{2}{\delta} \check{\boldsymbol{A}}_K \check{\boldsymbol{Q}}^{-1} \check{\boldsymbol{A}}_K^{\mathrm{T}} + \check{\boldsymbol{L}} \check{\boldsymbol{N}}^{-1} \check{\boldsymbol{L}}^{\mathrm{T}}\Big) \check{\boldsymbol{P}}\Big) \check{\boldsymbol{B}}$. The first inequality holds due to (4.21) and Assumption 4.3.4. The second inequality is obtained by eliminating the cross-terms by applying (A.2). Choosing $\nu > 0$ such that $\boldsymbol{M} - \frac{1}{\nu}\boldsymbol{\Upsilon} \succeq \boldsymbol{0}$, one obtains (B.24) with $c_3 = \frac{\delta}{2} \lambda^{\mathrm{m}}\left(\check{\boldsymbol{Q}}\right) > 0$.

In the second part it is shown that

$$c_1 \left\|\boldsymbol{\phi}\right\|^2 \leq V_N^*\left(\boldsymbol{\phi}\right) \leq c_2 \left\|\boldsymbol{\phi}\right\|^2 \quad \forall \boldsymbol{\phi} \in \Phi_N \tag{B.28}$$

where $c_1, c_2 > 0$. The lower bound is

$$V_N^*\left(\boldsymbol{\phi}\right) \geq W\left(\boldsymbol{\phi}\right) \geq c_1 \left\|\boldsymbol{\phi}\right\|^2 \quad \forall \boldsymbol{\phi} \in \Phi_N \tag{B.29}$$

with $c_1 = \lambda^{\mathrm{m}}\left(\check{\boldsymbol{P}}\right) > 0$.

The next step deals with the derivation of an upper bound on V_N^*. It is a modified version of the proof of [RM09, Proposition 2.18]. First note that there exists a parameter a such that

$$V_N^*\left(\boldsymbol{\phi}\right) \leq \max_{\boldsymbol{\phi} \in \Phi_N} \left(V_N^*\left(\boldsymbol{\phi}\right)\right) =: a. \tag{B.30}$$

The parameter always exists as $V_N^*\left(\boldsymbol{\phi}\right)$ is the sum of the continuous function $W\left(\boldsymbol{\phi}\right)$ and the function $\nu J_N^*\left(\hat{\boldsymbol{x}}\right)$ which can be bounded by the continuous function $J_N\left(\bar{\boldsymbol{V}}\right)$ as well as $\mathcal{X}$ and $\mathcal{U}$ are compact.

$\boldsymbol{\phi} \in \Phi_K$ implies that $J_N^*\left(\hat{\boldsymbol{x}}\right) = 0$ which leads to

$$V_N^*\left(\boldsymbol{\phi}\right) = W\left(\boldsymbol{\phi}\right) \leq \alpha\left(\left\|\boldsymbol{\phi}\right\|\right) \forall \boldsymbol{\phi} \in \Phi_K \tag{B.31}$$

with $\alpha(\|\boldsymbol{\phi}\|) = \lambda^{\mathrm{M}}(\check{\boldsymbol{P}})\|\boldsymbol{\phi}\|^2$.

Since the set Φ_K is non-empty (Lemma 4.3.3(a)), there exists an $\epsilon > 0$ such that $\mathcal{B}_\epsilon(\mathbf{0}) \subseteq \Phi_K$. Then one compute an upper bound $b = \max_{\boldsymbol{\phi}\in\mathcal{B}_\epsilon(\mathbf{0})}(\alpha(\|\boldsymbol{\phi}\|))$ where the solution exists due to the Weierstrass theorem. Since α is a strictly increasing, it follows that $b \leq \alpha(\|\boldsymbol{\phi}\|)\ \forall \boldsymbol{\phi} \in \Phi_N \setminus \Phi_K$ and thus

$$V_N^*(\boldsymbol{\phi}) \leq a \leq \frac{a}{b}\cdot\alpha(\|\boldsymbol{\phi}\|) \quad \forall \boldsymbol{\phi} \in \Phi_N \setminus \Phi_K. \tag{B.32}$$

Combining (B.31) and (B.32) leads to

$$V_N^*(\boldsymbol{\phi}) \leq c_2\|\boldsymbol{\phi}\|^2 \quad \forall \boldsymbol{\phi} \in \Phi_N \tag{B.33}$$

with $c_2 = \max\left(1, \frac{a}{b}\right)\cdot\lambda^{\mathrm{M}}(\check{\boldsymbol{P}})$. Hence, V_N^* is a Lyapunov function for the closed-loop system (4.18) and (4.19).

Using Lyapunov's theory (cf. Theorem 2.2.11) it follows that

$$\|\boldsymbol{\phi}_t\| \leq c\gamma^t\|\boldsymbol{\phi}_0\| \quad \forall \boldsymbol{\phi}_0 \in \Phi_N \tag{B.34}$$

for all $t \in \mathbb{N}$ with $c > 0$ and $0 < \gamma < 1$. This means that the origin of the closed-loop system (4.18) and (4.19) is GES with ROA Φ_N. This completes the proof. □

Proof of Theorem 4.4.2

Proof. The proof that $\tilde{\boldsymbol{x}}_t \in \Omega\ \forall t \in \mathbb{N}$ is identical to the proof of Theorem 4.3.1. The fact that $\hat{\boldsymbol{x}}_t \in \mathcal{X}_N$ and $\boldsymbol{u}_t \in \mathcal{U}\ \forall t \geq 0$ is proven in [BHA17, Theorem 1]. The argumentation that $\hat{\boldsymbol{x}}_t \in \mathcal{X} \ominus \Omega$ and $\boldsymbol{x}_t \in \mathcal{X}\ \forall t \in \mathbb{N}$ is again identical to the one in Theorem 4.3.1. This completes the proof. □

Proof of Theorem 4.4.3

Proof. Due to the absence of communication between the subsystems, the decentralized controllers can run asynchronously in the sense that they trigger at different time instants. Therefore, the optimal value function V_N^* might not exist at all times t and ideas from suboptimal MPC presented in Section 3.4 are used for analysis.

The closed-loop can be recast as

$$\boldsymbol{\phi}_{t+1} = \check{\boldsymbol{A}}_K\boldsymbol{\phi}_t + \check{\boldsymbol{B}}\boldsymbol{v}_t + \check{\boldsymbol{L}}\boldsymbol{e}_t - \check{\boldsymbol{B}}\boldsymbol{K}\boldsymbol{g}_t \tag{B.35}$$

$$\boldsymbol{V}_{t+1} = \boldsymbol{h}(\boldsymbol{V}_t, \boldsymbol{g}_t, \boldsymbol{\phi}_t) \tag{B.36}$$

where $\boldsymbol{e}_t^{\mathrm{T}} \boldsymbol{S}^{\mathrm{s}} \boldsymbol{e}_t \leq \boldsymbol{\phi}_t^{\mathrm{T}} \check{\boldsymbol{S}}^{\mathrm{s}} \boldsymbol{\phi}_t$, $\boldsymbol{g}_t^{\mathrm{T}} \boldsymbol{S}^{\mathrm{c}} \boldsymbol{g}_t \leq \boldsymbol{\phi}_t^{\mathrm{T}} \check{\boldsymbol{S}}^{\mathrm{c}} \boldsymbol{\phi}_t$ $\forall t \in \mathbb{N}$ and $\boldsymbol{v}_t = \mathrm{col}_{i \in \mathbb{Z}_{1:M}} \left(\bar{\boldsymbol{v}}^*_{i,t|t_{i,r}} \left(\hat{\boldsymbol{x}}_{i,t_{i,r}} \right) \right)$. The event-triggered computation of the new input sequence (4.24) is collected in the mapping $\boldsymbol{h}$. It is shown that the function $V_N(\boldsymbol{\phi}_t, \boldsymbol{V}_t) = V^{\mathrm{f}}(\boldsymbol{\phi}_t) + \nu J_N(\boldsymbol{V}_t)$ where $J_N(\boldsymbol{V}_t) = \sum_{i=1}^{M} J_N(\boldsymbol{V}_{i,t})$ is a Lyapunov function for the closed-loop system. Using Assumption 4.4.1 one can show in a similar way than done in Theorem 4.3.7 that for the cost difference

$$V_N(\boldsymbol{\phi}_{t+1}, \boldsymbol{V}_{t+1}) - V_N(\boldsymbol{\phi}_t, \boldsymbol{V}_t) \leq -c_6 \left\| (\boldsymbol{\phi}_t, \boldsymbol{v}_t) \right\|^2 \tag{B.37}$$

holds $\forall t \in \mathbb{N}$ where $c_6 = \min\left(\frac{\delta}{2}\lambda^{\mathrm{m}}\left(\check{\boldsymbol{Q}}\right), \lambda^{\mathrm{m}}\left(\boldsymbol{M} - \frac{1}{\nu}\boldsymbol{\Upsilon}\right)\right)$ with $\boldsymbol{M} - \frac{1}{\nu}\boldsymbol{\Upsilon} \succ \boldsymbol{0}$.

Now Theorem 3.4.2 which established stability of suboptimal MPC is applied. Above it has been shown that (3.9) holds.

Since $\boldsymbol{\phi} \in \Phi_K$ implies that $\boldsymbol{V} = \boldsymbol{0}$ and the set Φ_K is non-empty, there is a $\mathcal{B}_\epsilon(\boldsymbol{0}) \subseteq \Phi_K$ and $d \in \mathbb{R}_{>0}$ for which (3.10) holds. Note that in the considered case $\mathcal{Z} = \{\boldsymbol{0}\}$.

Consequently, there exist $\alpha > 0$ and $0 < \gamma < 1$ such that

$$\left\| (\boldsymbol{\phi}_t, \boldsymbol{V}_t) \right\| \leq \alpha \gamma^t \left\| (\boldsymbol{\phi}_0, \boldsymbol{V}_0) \right\| \tag{B.38}$$

for all $t \in \mathbb{N}$ and $\tilde{\alpha} > 0$ such that

$$\left\| \boldsymbol{\phi}_t \right\| \leq \tilde{\alpha} \gamma^t \left\| \boldsymbol{\phi}_0 \right\| \quad \forall \boldsymbol{\phi}_0 \in \Phi_N \tag{B.39}$$

for all $t \in \mathbb{N}$. This completes the proof. □

B.3 Chapter 5

Proof of Theorem 5.3.1

Proof. Assume that $\boldsymbol{x}_t \in \mathcal{X}_N$ and $\check{\boldsymbol{V}}_t \in \check{\mathcal{D}}_N(\boldsymbol{x}_t)$ are given as well as feasible sequences $\boldsymbol{V}_{i,t}$ for $i \in \mathbb{Z}_{1:M}$ are known. To prove recursive feasibility, it is shown that one can construct feasible sequences $\boldsymbol{V}_{i,t+1}$ from that. This is done in the following by outlining that the sequence obtained from shifting the old sequence and extending with $\boldsymbol{v}_{i,t+N|t+1} = \boldsymbol{0}$, i.e., $\boldsymbol{V}_{i,t+1} = \left[\boldsymbol{v}_{i,t+1|t}^{\mathrm{T}}, \dots, \boldsymbol{v}_{i,t+N-1|t}^{\mathrm{T}}, \boldsymbol{0}\right]^{\mathrm{T}}$, satisfies the constraints of optimization problem (5.18).

The deviation between the sequences starting from $\boldsymbol{x}_{i,t+1}$ and $\bar{\boldsymbol{x}}_{i,t+1|t}$ as well as $\check{\boldsymbol{x}}_{j,t+1}$ and $\check{\boldsymbol{x}}_{j,t+1|t}$ (constraint (5.18b)) for $j \in \mathcal{N}_i$ can be written as

$$\check{\boldsymbol{x}}_{j,t+r+1|t+1} - \check{\boldsymbol{x}}_{j,t+r+1|t} = \boldsymbol{F}_{jj}\left(\check{\boldsymbol{x}}_{j,t+r|t+1} - \check{\boldsymbol{x}}_{j,t+r|t}\right) \tag{B.40}$$

$$\in \boldsymbol{F}_{jj}\check{\mathcal{W}}_j^r =: \check{\mathcal{W}}_j^{r+1} \tag{B.41}$$

$$\bar{\boldsymbol{x}}_{i,t+r+1|t+1} - \bar{\boldsymbol{x}}_{i,t+r+1|t} = \boldsymbol{F}_{ii}\left(\bar{\boldsymbol{x}}_{i,t+r|t+1} - \bar{\boldsymbol{x}}_{i,t+r|t}\right) + \sum_{j\in\mathcal{N}_i} \boldsymbol{F}_{ij}\left(\check{\boldsymbol{x}}_{j,t+r|t+1} - \check{\boldsymbol{x}}_{j,t+r|t}\right) \tag{B.42}$$

$$\in \boldsymbol{F}_{ii}\bar{\mathcal{W}}_i^r \oplus \bigoplus_{j\in\mathcal{N}_i} \boldsymbol{F}_{ij}\check{\mathcal{W}}_j^r =: \bar{\mathcal{W}}_i^{r+1} \tag{B.43}$$

for $r \in \mathbb{Z}_{0:N-1}$. For the input follows

$$\bar{\boldsymbol{u}}_{i,t+r|t+1} - \bar{\boldsymbol{u}}_{i,t+r|t} = \boldsymbol{K}_{ii}\left(\bar{\boldsymbol{x}}_{i,t+r|t+1} - \bar{\boldsymbol{x}}_{i,t+r|t}\right) + \sum_{j\in\mathcal{N}_i} \boldsymbol{K}_{ij}\left(\check{\boldsymbol{x}}_{j,t+r|t+1} - \check{\boldsymbol{x}}_{j,t+r|t}\right) \tag{B.44}$$

$$\in \boldsymbol{K}_{ii}\bar{\mathcal{W}}_i^r \oplus \bigoplus_{j\in\mathcal{N}_i} \boldsymbol{K}_{ij}\check{\mathcal{W}}_j^r =: \bar{\mathcal{W}}_{\mathrm{u},i}^r. \tag{B.45}$$

for $r \in \mathbb{Z}_{1:N-1}$. This means that

$$\bar{\boldsymbol{x}}_{i,t+r|t+1} \in \left\{\bar{\boldsymbol{x}}_{i,t+r|t}\right\} \oplus \bar{\mathcal{W}}_i^r \subseteq \bar{\mathcal{X}}_i^r \oplus \bar{\mathcal{W}}_i^r \subseteq \bar{\mathcal{X}}_i^{r-1} \quad \forall r \in \mathbb{Z}_{1:N}, \tag{B.46}$$

$$\bar{\boldsymbol{u}}_{i,t+r|t+1} \in \left\{\bar{\boldsymbol{u}}_{i,t+r|t}\right\} \oplus \bar{\mathcal{W}}_{\mathrm{u},i}^r \subseteq \bar{\mathcal{U}}_i^r \oplus \bar{\mathcal{W}}_{\mathrm{u},i}^r \subseteq \bar{\mathcal{U}}_i^{r-1} \quad \forall r \in \mathbb{Z}_{1:N-1}. \tag{B.47}$$

which implies that the constraints (5.18f) are feasible for $t+1$.

For the states at the end of the prediction horizon $\check{\boldsymbol{x}}_{j,t+N|t+1} \in \left\{\check{\boldsymbol{x}}_{j,t+N|t}\right\} \oplus \check{\mathcal{W}}_j^N \subseteq \check{\bar{\mathcal{X}}}_j^{\mathrm{f}} \oplus \check{\mathcal{W}}_j^N$ and $\bar{\boldsymbol{x}}_{i,t+N|t+1} \in \left\{\bar{\boldsymbol{x}}_{i,t+N|t}\right\} \oplus \bar{\mathcal{W}}_i^N \subseteq \bar{\mathcal{X}}_i^{\mathrm{f}} \oplus \bar{\mathcal{W}}_i^N$ holds. As the sets $\bar{\mathcal{X}}_i^{\mathrm{f}}$ and $\check{\bar{\mathcal{X}}}_j^{\mathrm{f}}$ fulfill Assumption 5.2.2, it follows that $\bar{\boldsymbol{x}}_{i,t+N+1|t+1} \in \bar{\mathcal{X}}_i^{\mathrm{f}}$ and $\check{\boldsymbol{x}}_{j,t+N+1|t+1} \in \check{\bar{\mathcal{X}}}_j^{\mathrm{f}}$. Thus, constraint (5.18c) is feasible. Moreover, Assumption 5.2.2 implies that $\bar{\boldsymbol{u}}_{i,t+N|t+1} \in \bar{\mathcal{U}}_i^{N-1}$ which means that the constraints (5.18g) are feasible for $t+1$.

What is left to show is feasibility of (5.18h). For $r \in \mathbb{Z}_{1:N-1}$ (5.18h) is feasible since the sequence is shifted and satisfies the constraints at t. For step $r = N$ (5.18h) is feasible as $\boldsymbol{v}_{i,t+N|t+1} = \mathbf{0}$ and $\check{\boldsymbol{v}}_{i,t+N|t+1} = \mathbf{0}$ and thus $\boldsymbol{v}_{i,t+N|t+1} - \check{\boldsymbol{v}}_{i,t+N|t+1} = \mathbf{0} \in \mathcal{V}_i$.

This means $\check{\boldsymbol{V}}_{t+1} \in \check{\mathcal{D}}_N(\boldsymbol{x}_{t+1})$ and $\boldsymbol{x}_{t+1} \in \mathcal{X}_N$. From $\boldsymbol{x}_0 \in \mathcal{X}_N$ and $\check{\boldsymbol{V}}_0 \in \check{\mathcal{D}}_N(\boldsymbol{x}_0)$ follows that the problem is initially feasible and thus $\boldsymbol{x}_t \in \mathcal{X}_N$, $\check{\boldsymbol{V}}_t \in \check{\mathcal{D}}_N(\boldsymbol{x}_t)$ and $\boldsymbol{V}_t \in \mathcal{D}_N\left(\boldsymbol{x}_t, \check{\boldsymbol{V}}_t\right)$, $t \in \mathbb{N}$. This completes the proof. □

Proof of Theorem 5.3.5

Proof. It follows from Theorem 5.3.1 that the control law is well-defined $\forall t \in \mathbb{N}$. Moreover, it can be derived in a similar way than done in Lemma 2.3.3 that

$$J_N^*\left(\boldsymbol{x}_{t+1}, \check{\boldsymbol{V}}_t\right) - J_N^*\left(\boldsymbol{x}_t, \check{\boldsymbol{V}}_t\right) \leq -\boldsymbol{v}_t^{\mathrm{T}}\boldsymbol{M}\boldsymbol{v}_t \tag{B.48}$$

$\forall t \in \mathbb{N}$ holds. It follows that $J_N^*\left(\boldsymbol{x}_t, \check{\boldsymbol{V}}_t\right)$ is a non-increasing sequence. Moreover, it is bounded below by zero. Thus, $\lim_{t\to\infty} J_N^*\left(\boldsymbol{x}_t, \check{\boldsymbol{V}}_t\right) = J^{\mathrm{fp}} \geq 0$. Thus, the left hand side of (B.48) goes to zero which implies together with $\boldsymbol{M} \succ \mathbf{0}$ that $\lim\limits_{t\to\infty} \|\boldsymbol{v}_t\| = 0$.

Since $\check{\boldsymbol{V}}_t$ is updated as described in step 5 of Algorithm 1, i.e., by shifting the previous sequence, removing the first value and extending it with zeros, it follows that $\lim\limits_{t\to\infty} \left\|\check{\boldsymbol{V}}_t\right\| = 0$. Now consider the difference

$$\begin{aligned}
&V_N^*\left(\boldsymbol{x}_{t+1}, \check{\boldsymbol{V}}_{t+1}\right) - V_N^*\left(\boldsymbol{x}_t, \check{\boldsymbol{V}}_t\right) && \text{(B.49)}\\
=&V^{\mathrm{f}}(\boldsymbol{x}_{t+1}) + \nu J_N^*\left(\boldsymbol{x}_{t+1}, \check{\boldsymbol{V}}_{t+1}\right) - V^{\mathrm{f}}(\boldsymbol{x}_t) - \nu J_N^*\left(\boldsymbol{x}_t, \check{\boldsymbol{V}}_t\right) && \text{(B.50)}\\
\leq&\left(\boldsymbol{F}\boldsymbol{x}_t + \boldsymbol{B}\boldsymbol{K}_{\mathrm{C}}\boldsymbol{e}_t + \boldsymbol{B}\boldsymbol{v}_t\right)^{\mathrm{T}}\boldsymbol{P}\left(\boldsymbol{F}\boldsymbol{x}_t + \boldsymbol{B}\boldsymbol{K}_{\mathrm{C}}\boldsymbol{e}_t + \boldsymbol{B}\boldsymbol{v}_t\right) - \boldsymbol{x}_t^{\mathrm{T}}\boldsymbol{P}\boldsymbol{x}_t - \nu\boldsymbol{v}_t^{\mathrm{T}}\boldsymbol{M}\boldsymbol{v}_t && \text{(B.51)}\\
\leq&-\delta\boldsymbol{x}_t^{\mathrm{T}}\boldsymbol{Q}_K\boldsymbol{x}_t + \boldsymbol{v}_t^{\mathrm{T}}\boldsymbol{B}^{\mathrm{T}}\boldsymbol{P}\boldsymbol{B}\boldsymbol{v}_t + 2\boldsymbol{v}_t^{\mathrm{T}}\boldsymbol{B}^{\mathrm{T}}\boldsymbol{P}\boldsymbol{F}\boldsymbol{x}_t + 2\boldsymbol{e}_t^{\mathrm{T}}\boldsymbol{K}_{\mathrm{C}}^{\mathrm{T}}\boldsymbol{B}^{\mathrm{T}}\boldsymbol{P}\boldsymbol{B}\boldsymbol{v}_t - \boldsymbol{e}_t^{\mathrm{T}}\boldsymbol{N}\boldsymbol{e}_t - \nu\boldsymbol{v}_t^{\mathrm{T}}\boldsymbol{M}\boldsymbol{v}_t\\
& && \text{(B.52)}\\
\leq&-\frac{\delta}{2}\boldsymbol{x}_t^{\mathrm{T}}\boldsymbol{Q}_K\boldsymbol{x}_t - \boldsymbol{v}_t^{\mathrm{T}}\left(\nu\boldsymbol{M} - \boldsymbol{\Upsilon}\right)\boldsymbol{v}_t && \text{(B.53)}
\end{aligned}$$

where $\boldsymbol{\Upsilon} = \boldsymbol{B}^{\mathrm{T}}\left(\boldsymbol{P} + \boldsymbol{P}\left(\boldsymbol{B}\boldsymbol{K}_{\mathrm{C}}\boldsymbol{N}^{-1}\boldsymbol{K}_{\mathrm{C}}^{\mathrm{T}}\boldsymbol{B}^{\mathrm{T}} + \frac{2}{\delta}\boldsymbol{F}\boldsymbol{Q}_K^{-1}\boldsymbol{F}^{\mathrm{T}}\right)\boldsymbol{P}\right)\boldsymbol{B}$. For obtaining the first inequality, (B.48) is used and (5.29) is substituted. The second inequality follows from Assumption 5.3.2 and for the third (A.2) is applied. Choosing $\boldsymbol{M} \succ \frac{1}{\nu}\boldsymbol{\Upsilon}$, it follows that

$$V_N^*\left(\boldsymbol{x}_{t+1}, \check{\boldsymbol{V}}_{t+1}\right) - V_N^*\left(\boldsymbol{x}_t, \check{\boldsymbol{V}}_t\right) \leq -\frac{\delta}{2}\boldsymbol{x}_t^{\mathrm{T}}\boldsymbol{Q}_K\boldsymbol{x}_t. \tag{B.54}$$

This means that $V_N^*(\boldsymbol{x}_t, \boldsymbol{V}_t)$ is a non-increasing sequence. Furthermore, it is bounded below by zero. Thus, $\lim_{t\to\infty} V_N^*(\boldsymbol{x}_t, \boldsymbol{V}_t) = V^{\mathrm{fp}} \geq 0$. Thus, the left hand side of (B.54) goes to zero which implies together with $\boldsymbol{Q}_K \succ \mathbf{0}$ that $\lim\limits_{t\to\infty} \|\boldsymbol{x}_t\| = 0$.

Since $\lim_{t\to\infty} \|\boldsymbol{x}_t\| = 0$ and $\lim_{t\to\infty} \|\boldsymbol{v}_t\| = 0$, it follows that $\lim_{t\to\infty} \|\boldsymbol{u}_t\| = 0$. This completes the proof. □

Proof of Lemma 5.3.3

Proof. Multiplying the matrix inequality (5.31) from right and left with $\left[\boldsymbol{x}^{\mathrm{T}}, \boldsymbol{e}^{\mathrm{T}}\right]^{\mathrm{T}}$ and its transpose, respectively, one obtains

$$\boldsymbol{x}_t^{\mathrm{T}}\left(\boldsymbol{P}-\boldsymbol{F}^{\mathrm{T}}\boldsymbol{P}\boldsymbol{F}-\delta\boldsymbol{Q}_K-\sigma\boldsymbol{S}\right)\boldsymbol{x}_t+2\boldsymbol{x}_t^{\mathrm{T}}\boldsymbol{F}^{\mathrm{T}}\boldsymbol{P}\boldsymbol{B}\boldsymbol{K}_{\mathrm{C}}\boldsymbol{e}_t+\boldsymbol{e}_t^{\mathrm{T}}\left(\boldsymbol{S}-\boldsymbol{K}_{\mathrm{C}}^{\mathrm{T}}\boldsymbol{B}^{\mathrm{T}}\boldsymbol{P}\boldsymbol{B}\boldsymbol{K}_{\mathrm{C}}-\boldsymbol{N}\right)\boldsymbol{e}_t\geq 0 \tag{B.55}$$

Applying (5.30), it follows that (5.32) holds. This completes the proof. □

Proof of Lemma 5.3.4

Proof. Since $\boldsymbol{F}$ is stable, there exists a matrix $\boldsymbol{P} \succ \boldsymbol{0}$ for $\boldsymbol{Q}_K \succ \boldsymbol{0}$ such that $\boldsymbol{F}^{\mathrm{T}}\boldsymbol{P}\boldsymbol{F}-\boldsymbol{P} = -\boldsymbol{Q}_K$. Evaluating the difference of the function V^{f} along $\boldsymbol{x}_{t+1} = \boldsymbol{F}\boldsymbol{x}_t+\boldsymbol{B}\boldsymbol{K}_{\mathrm{C}}\boldsymbol{e}_t$, it follows that

$$\begin{aligned} V^{\mathrm{f}}(\boldsymbol{x}_{t+1}) - V^{\mathrm{f}}(\boldsymbol{x}_t) &= \left(\boldsymbol{F}\boldsymbol{x}_t + \boldsymbol{B}\boldsymbol{K}_{\mathrm{C}}\boldsymbol{e}_t\right)^{\mathrm{T}} \boldsymbol{P}\left(\boldsymbol{F}\boldsymbol{x}_t + \boldsymbol{B}\boldsymbol{K}_{\mathrm{C}}\boldsymbol{e}_t\right) - \boldsymbol{x}_t^{\mathrm{T}}\boldsymbol{P}\boldsymbol{x}_t & \text{(B.56)} \\ &= -\boldsymbol{x}_t^{\mathrm{T}}\boldsymbol{Q}_K\boldsymbol{x}_t + 2\boldsymbol{x}_t^{\mathrm{T}}\boldsymbol{F}^{\mathrm{T}}\boldsymbol{P}\boldsymbol{B}\boldsymbol{K}_{\mathrm{C}}\boldsymbol{e}_t + \boldsymbol{e}_t^{\mathrm{T}}\boldsymbol{K}_{\mathrm{C}}\boldsymbol{B}^{\mathrm{T}}\boldsymbol{P}\boldsymbol{B}\boldsymbol{K}_{\mathrm{C}}\boldsymbol{e}_t & \text{(B.57)} \\ &\leq (\lambda-1)\boldsymbol{x}_t^{\mathrm{T}}\boldsymbol{Q}_K\boldsymbol{x}_t + \boldsymbol{e}_t^{\mathrm{T}}\boldsymbol{\Psi}\boldsymbol{e}_t - \boldsymbol{e}_t^{\mathrm{T}}\boldsymbol{N}\boldsymbol{e}_t + \left(-\boldsymbol{e}_t^{\mathrm{T}}\boldsymbol{S}\boldsymbol{e}_t + \sigma\boldsymbol{x}_t^{\mathrm{T}}\boldsymbol{S}\boldsymbol{x}_t\right) & \text{(B.58)} \end{aligned}$$

where for $\boldsymbol{\Psi} = \boldsymbol{N} + \boldsymbol{K}_{\mathrm{C}}^{\mathrm{T}}\boldsymbol{B}^{\mathrm{T}}\left(\lambda^{-1}\boldsymbol{P}\boldsymbol{F}\boldsymbol{Q}_K^{-1}\boldsymbol{F}^{\mathrm{T}}\boldsymbol{P} + \boldsymbol{P}\right)\boldsymbol{B}\boldsymbol{K}_{\mathrm{C}}$ with $\lambda > 0$. Note that for the second step (A.2) is applied. Choosing $\boldsymbol{S} = \lambda^{\mathrm{M}}(\boldsymbol{\Psi}) \cdot \boldsymbol{I} \succ \boldsymbol{0}$ and $0 \leq \sigma \leq \bar{\sigma}$ where $\bar{\sigma} = \frac{(1-\delta-\lambda)\lambda^{\mathrm{m}}(\boldsymbol{Q}_K)}{2\lambda^{\mathrm{M}}(\boldsymbol{S})} > 0$ implies that (5.32) holds. For this purpose δ and λ have to be chosen such that $1-\delta-\lambda > 0$. This completes the proof. □

Proof of Lemma 5.4.1

Proof. Assume the initialization problems (5.33) are feasible and let $\check{\check{\boldsymbol{V}}}_i^*$ be the minimizers for all $t \in \mathbb{Z}_{1:M}$. $\check{\check{\boldsymbol{x}}}_{i,r}$ for $r \in \mathbb{Z}_N$ and $\check{\check{\boldsymbol{u}}}_{i,r}$ for $r \in \mathbb{Z}_{N-1}$ are the corresponding state and input sequences, respectively. Moreover, let $\bar{\boldsymbol{x}}_{i,r}$ for $r \in \mathbb{Z}_N$ and $\bar{\boldsymbol{u}}_{i,r}$ for $r \in \mathbb{Z}_{N-1}$ be the sequences constructed using the model in (5.18) with $\boldsymbol{V}_i = \check{\check{\boldsymbol{V}}}_i^*$ and $\check{\boldsymbol{V}}_j = \check{\check{\boldsymbol{V}}}_j^*$ for all $j \in \mathcal{N}_i$ and $i \in \mathbb{Z}_{1:M}$. To show that feasibility of (5.33) implies feasibility of (5.18), it has to be shown that $\check{\check{\boldsymbol{x}}}_{i,r} \in \check{\check{\mathcal{X}}}_i^r \Rightarrow \bar{\boldsymbol{x}}_{i,r} \in \bar{\mathcal{X}}_i^r$, $\check{\check{\boldsymbol{u}}}_{i,r} \in \check{\check{\mathcal{U}}}_i^r \Rightarrow \bar{\boldsymbol{u}}_{i,r} \in \bar{\mathcal{U}}_i^r$ for all $r \in \mathbb{Z}_{0:N-1}$, $\check{\check{\boldsymbol{x}}}_{i,N} \in \check{\check{\mathcal{X}}}_i^{\mathrm{f}} \Rightarrow \bar{\boldsymbol{x}}_{i,N} \in \bar{\mathcal{X}}_i^{\mathrm{f}}$ and $\check{\check{\boldsymbol{x}}}_{i,N} \in \check{\check{\mathcal{X}}}_i^{\mathrm{f}} \Rightarrow \check{\boldsymbol{x}}_{i,N} \in \check{\mathcal{X}}_i^{\mathrm{f}}$.

For the difference between the states $\bar{\boldsymbol{x}}_{i,r+1}$ and $\check{\check{\boldsymbol{x}}}_{i,r+1}$ it follows for $r \in \mathbb{Z}_{0:N-1}$

$$\bar{\boldsymbol{x}}_{i,r+1} - \check{\check{\boldsymbol{x}}}_{i,r+1} = \boldsymbol{F}_{ii}\left(\bar{\boldsymbol{x}}_{i,r} - \check{\check{\boldsymbol{x}}}_{i,r}\right) + \sum_{j \in \mathcal{N}_i} \boldsymbol{F}_{ij}\check{\boldsymbol{x}}_{j,r} \tag{B.59}$$

$$\in \boldsymbol{F}_{ii}\check{\check{\mathcal{W}}}_i^r \oplus \bigoplus_{j \in \mathcal{N}_i} \boldsymbol{F}_{ij}\bar{\mathcal{X}}_j^r \tag{B.60}$$

$$=: \check{\check{\mathcal{W}}}_i^{r+1} \tag{B.61}$$

which leads to

$$\bar{\boldsymbol{x}}_{i,r} \in \left\{\check{\check{\boldsymbol{x}}}_{i,r}\right\} \oplus \check{\check{\mathcal{W}}}_i^r \subseteq \check{\check{\mathcal{X}}}_i^r \oplus \check{\check{\mathcal{W}}}_i^r \subseteq \bar{\mathcal{X}}_i^r \quad \forall r \in \mathbb{Z}_{0:N-1} \tag{B.62}$$

$$\bar{\boldsymbol{x}}_{i,N} \in \left\{\check{\check{\boldsymbol{x}}}_{i,N}\right\} \oplus \check{\check{\mathcal{W}}}_i^N \subseteq \check{\check{\mathcal{X}}}_i^{\mathrm{f}} \oplus \check{\check{\mathcal{W}}}_i^N \subseteq \bar{\mathcal{X}}_i^{\mathrm{f}}. \tag{B.63}$$

As $\check{\check{\boldsymbol{x}}}_{i,r} = \check{\boldsymbol{x}}_{i,r}$ for all $r \in \mathbb{Z}_{0:N}$, it follows that $\check{\boldsymbol{x}}_{i,N} = \check{\check{\boldsymbol{x}}}_{i,N} \in \check{\check{\mathcal{X}}}_i^{\mathrm{f}} \subseteq \bar{\check{\mathcal{X}}}_i^{\mathrm{f}}$.

For the input

$$\bar{\boldsymbol{u}}_{i,r} - \check{\check{\boldsymbol{u}}}_{i,r} = \boldsymbol{K}_{ii}\left(\bar{\boldsymbol{x}}_{i,r} - \check{\check{\boldsymbol{x}}}_{i,r}\right) + \sum_{j \in \mathcal{N}_i} \boldsymbol{K}_{ij}\check{\boldsymbol{x}}_{j,r} \tag{B.64}$$

$$\in \boldsymbol{K}_{ii}\check{\check{\mathcal{W}}}_i^r \oplus \bigoplus_{j \in \mathcal{N}_i} \boldsymbol{K}_{ij}\bar{\mathcal{X}}_j^r \tag{B.65}$$

holds for $r \in \mathbb{Z}_{0:N-1}$. This leads to

$$\bar{\boldsymbol{u}}_{i,r} \in \left\{\check{\check{\boldsymbol{u}}}_{i,r}\right\} \oplus \boldsymbol{K}_{ii}\check{\check{\mathcal{W}}}_i^r \oplus \bigoplus_{j \in \mathcal{N}_i} \boldsymbol{K}_{ij}\bar{\mathcal{X}}_j^r \tag{B.66}$$

$$\subseteq \check{\check{\mathcal{U}}}_i^r \oplus \boldsymbol{K}_{ii}\check{\check{\mathcal{W}}}_i^r \oplus \bigoplus_{j \in \mathcal{N}_i} \boldsymbol{K}_{ij}\bar{\mathcal{X}}_j^r \subseteq \bar{\mathcal{U}}_i^r \tag{B.67}$$

for $r \in \mathbb{Z}_{0:N-1}$. This completes the proof. □

B.4 Chapter 6

Proof of Lemma 6.3.1

Proof. Recall that the predictions at time $\tilde{t} \in \{t, t+1\}$ starting from initial states $\bar{\boldsymbol{x}}_{i,t+1|t}$ and $\bar{\boldsymbol{x}}_{i,t+1|t+1} = \bar{\boldsymbol{x}}_{i,t+1}$, respectively, using the sequence $\boldsymbol{V}_{i,t+1}$ satisfy

$$\bar{\boldsymbol{x}}_{i,r+1|\tilde{t}} = \boldsymbol{F}_{ii}\bar{\boldsymbol{x}}_{i,r|\tilde{t}} + \boldsymbol{B}_{ii}\boldsymbol{v}_{i,r|t} + \sum_{j\in\mathcal{C}_{i,\tilde{t}}} \boldsymbol{F}_{ij}\bar{\boldsymbol{x}}_{j,r|\tilde{t}} + \sum_{j\in\mathcal{N}_i\backslash\mathcal{C}_{i,\tilde{t}}} \boldsymbol{F}_{ij}\breve{\boldsymbol{x}}^{[i]}_{j,r|\tilde{t}} \tag{B.68}$$

$\forall r \in \mathbb{Z}_{\tilde{t}:t+N-1}$, $\forall i \in \mathbb{Z}_{1:M}$. Computing the difference for all $r \in \mathbb{Z}_{t:t+N-1}$ leads to

$$\begin{aligned}
\bar{\boldsymbol{x}}_{i,r+1|t+1} - \bar{\boldsymbol{x}}_{i,r+1|t} =& \boldsymbol{F}_{ii}\left(\bar{\boldsymbol{x}}_{i,r|t+1} - \bar{\boldsymbol{x}}_{i,r|t}\right) + \sum_{j\in\mathcal{C}_{i,t+1}\cap\mathcal{C}_{i,t}} \boldsymbol{F}_{ij}\left(\bar{\boldsymbol{x}}_{j,r|t+1} - \bar{\boldsymbol{x}}_{j,r|t}\right)\\
&+ \sum_{j\in\mathcal{C}_{i,t+1}\backslash\mathcal{C}_{i,t}} \boldsymbol{F}_{ij}\left(\bar{\boldsymbol{x}}_{j,r|t+1} - \breve{\boldsymbol{x}}^{[i]}_{j,r|t}\right) + \sum_{j\in\mathcal{C}_{i,t}\backslash\mathcal{C}_{i,t+1}} \boldsymbol{F}_{ij}\left(\breve{\boldsymbol{x}}^{[i]}_{j,r|t+1} - \bar{\boldsymbol{x}}_{j,r|t}\right)\\
&+ \sum_{j\in\mathcal{N}_i\backslash(\mathcal{C}_{i,t}\cup\mathcal{C}_{i,t+1})} \boldsymbol{F}_{ij}\left(\breve{\boldsymbol{x}}^{[i]}_{j,r|t+1} - \breve{\boldsymbol{x}}^{[i]}_{j,r|t}\right) &\text{(B.69)}\\
=& \boldsymbol{F}_{ii}\left(\bar{\boldsymbol{x}}_{i,r|t+1} - \bar{\boldsymbol{x}}_{i,r|t}\right) + \sum_{j\in\mathcal{C}_{i,t+1}\cap\mathcal{C}_{i,t}} \boldsymbol{F}_{ij}\left(\bar{\boldsymbol{x}}_{j,r|t+1} - \bar{\boldsymbol{x}}_{j,t+r|t}\right)\\
&+ \sum_{j\in\mathcal{C}_{i,t+1}\backslash\mathcal{C}_{i,t}} \boldsymbol{F}_{ij}\left(\bar{\boldsymbol{x}}_{j,r|t+1} - \bar{\boldsymbol{x}}_{j,r|t} + \bar{\boldsymbol{x}}_{j,r|t} - \breve{\boldsymbol{x}}^{[i]}_{j,r|t}\right) &\text{(B.70)}\\
=& \boldsymbol{F}_{ii}\left(\bar{\boldsymbol{x}}_{i,r|t+1} - \bar{\boldsymbol{x}}_{i,r|t}\right) + \sum_{j\in\mathcal{C}_{i,t+1}} \boldsymbol{F}_{ij}\left(\bar{\boldsymbol{x}}_{j,r|t+1} - \bar{\boldsymbol{x}}_{j,t+r|t}\right)\\
&+ \sum_{j\in\mathcal{C}_{i,t+1}\backslash\mathcal{C}_{i,t}} \boldsymbol{F}_{ij}\left(\bar{\boldsymbol{x}}_{j,r|t} - \breve{\boldsymbol{x}}^{[i]}_{j,r|t}\right). &\text{(B.71)}
\end{aligned}$$

The second equality holds since $\breve{\boldsymbol{x}}^{[i]}_{j,r|t+1} = \bar{\boldsymbol{x}}_{j,r|t}$ $\forall j \in \mathcal{C}_{i,t}\backslash\mathcal{C}_{i,t+1}$ $\forall r \in \mathbb{Z}_{t:t+N-1}$, $\breve{\boldsymbol{x}}^{[i]}_{j,r|t+1} = \breve{\boldsymbol{x}}^{[i]}_{j,r|t}$ $\forall j \in \mathcal{N}_i\backslash(\mathcal{C}_{i,t} \cup \mathcal{C}_{i,t+1})$ $\forall r \in \mathbb{Z}_{t:t+N-1}$. The third equality follows from rearranging the terms.

From $\bar{\boldsymbol{x}}_{i,t+1|t+1} - \bar{\boldsymbol{x}}_{i,t+1|t} = \boldsymbol{w}_t \in \mathcal{W}_i$ and (6.10), it follows that

$$\begin{aligned}
&\bar{\boldsymbol{x}}_{i,t+r+1|t+1} - \bar{\boldsymbol{x}}_{i,t+r+1|t} \in \mathcal{H}_i^{r+1} && \forall r \in \mathbb{Z}_{0:N-1} &\text{(B.72)}\\
&\mathcal{H}_i^{r+1} = \bigoplus_{j\in\mathcal{N}_i\cup\{i\}} \boldsymbol{F}_{ij}\mathcal{H}_j^{r} \oplus \mathcal{J}_i && \forall r \in \mathbb{Z}_{1:N-1} &\text{(B.73)}\\
&\mathcal{H}_i^{1} = \mathcal{W}_i \text{ and } \mathcal{J}_i = \bigoplus_{j\in\mathcal{N}_i} \boldsymbol{F}_{ij}\mathcal{E}_j^{\mathrm{x}} && \forall i \in \mathbb{Z}_{1:M}. &\text{(B.74)}
\end{aligned}$$

This completes the proof. □

Proof of Lemma 6.3.4

Proof. It is shown that the shifted sequence $\boldsymbol{V}_{t+1}^{\text{shift}} = \left[\boldsymbol{v}_{t+1|t}^{\text{T}}, \ldots, \boldsymbol{v}_{t+N-1|t}^{\text{T}}, \boldsymbol{0}\right]^{\text{T}}$ satisfies the constraints in (6.8) at time $t+1$.

First, it is proven that the state constraints (6.8h) are satisfied, i.e., $\bar{\boldsymbol{x}}_{t+r|t+1} \in \bar{\mathcal{X}}^{r-1} \forall r \in \mathbb{Z}_{1:N}$. From Lemma 6.3.1 and the state constraints in (6.8) it follows that $\bar{\boldsymbol{x}}_{t+r|t+1} \in \left\{\bar{\boldsymbol{x}}_{t+r|t}\right\} \oplus \mathcal{H}^r \subseteq \bar{\mathcal{X}}^r \oplus \mathcal{H}^r$ for all $r \in \mathbb{Z}_{0:N-1}$. Together with (6.16) this leads to

$$\bar{\boldsymbol{x}}_{t+r|t+1} \in \left(\bar{\mathcal{X}}^{r-1} \ominus \mathcal{H}^r\right) \oplus \mathcal{H}^r \subseteq \bar{\mathcal{X}}^{r-1} \quad \forall r \in \mathbb{Z}_{1:N-1}. \tag{B.75}$$

The same arguments together with Assumption 6.3.2 lead to $\bar{\boldsymbol{x}}_{t+N|t+1} \in \bar{\mathcal{X}}^{N-1}$.

In the next step, it has to be shown for the input constraints that $\bar{\boldsymbol{u}}_{i,t+r|t+1} \in \bar{\mathcal{U}}_i^{r-1} \forall r \in \mathbb{Z}_{1:N}$. For the input at time t holds

$$\bar{\boldsymbol{u}}_{i,t+r|t} = \boldsymbol{v}_{i,t+r|t} + \sum_{j \in \mathcal{C}_{i,t} \cup \{i\}} \boldsymbol{K}_{ij} \bar{\boldsymbol{x}}_{j,t+r|t} + \sum_{j \in \mathcal{N}_i \setminus \mathcal{C}_{i,t}} \boldsymbol{K}_{ij} \breve{\boldsymbol{x}}_{j,t+r|t}^{[i]} \tag{B.76}$$

for $r \in \mathbb{Z}_{0:N-1}$ and at $t+1$

$$\bar{\boldsymbol{u}}_{i,t+r|t+1} = \boldsymbol{v}_{i,t+r|t} + \sum_{j \in \mathcal{C}_{i,t+1} \cup \{i\}} \boldsymbol{K}_{ij} \bar{\boldsymbol{x}}_{j,t+r|t+1} + \sum_{j \in \mathcal{N}_i \setminus \mathcal{C}_{i,t+1}} \boldsymbol{K}_{ij} \breve{\boldsymbol{x}}_{j,t+r|t+1}^{[i]} \tag{B.77}$$

for $r \in \mathbb{Z}_{1:N-1}$, respectively. As $\bar{\boldsymbol{u}}_{i,t+r|t} \in \bar{\mathcal{U}}_i^r$

$$\bar{\boldsymbol{u}}_{i,t+r|t+1} \in (\bar{\mathcal{U}}_i^{r-1} \ominus \bigoplus_{j \in \mathcal{N}_i \cup \{i\}} \boldsymbol{K}_{ij} \mathcal{H}_j^r \ominus \bigoplus_{j \in \mathcal{N}_i} \boldsymbol{K}_{ij} \mathcal{E}_j^{\text{x}}) \oplus \bigoplus_{j \in \mathcal{N}_i \cup \{i\}} \boldsymbol{K}_{ij} \mathcal{H}_j^r \oplus \bigoplus_{j \in \mathcal{N}_i} \boldsymbol{K}_{ij} \mathcal{E}_j^{\text{x}} \tag{B.78}$$

and thus $\bar{\boldsymbol{u}}_{t+r|t+1} \in \bar{\mathcal{U}}^{r-1}$ $\forall r \in \mathbb{Z}_{1:N-1}$. For the last input

$$\bar{\boldsymbol{u}}_{i,t+N|t+1} = \sum_{j \in \mathcal{N}_i \cup \{i\}} \boldsymbol{K}_{ij} \bar{\boldsymbol{x}}_{i,t+N|t+1} \in \bar{\mathcal{U}}_i^{N-1} \tag{B.79}$$

holds (see Assumption 6.3.2). Consequently, the input constraints (6.8g) are satisfied.

Next, it is shown that $\bar{\boldsymbol{x}}_{i,t+N+1|t+1} \in \mathcal{X}_i^{\text{f}}(\alpha_{i,t+1})$ $\forall i \in \mathbb{Z}_{1:M}$. Since $\bar{\boldsymbol{x}}_t \in \mathcal{X}_N$, $\bar{\boldsymbol{x}}_{i,t+N|t} \in \mathcal{X}_i^{\text{f}}(\alpha_{i,t})$ holds which implies that $\bar{\boldsymbol{x}}_{t+N|t} \in \mathcal{X}_{\text{Glo}}^{\text{f}}$. Lemma 6.3.1 and Assumption 6.3.2 lead to $\bar{\boldsymbol{x}}_{t+N|t+1} \in \left\{\bar{\boldsymbol{x}}_{t+N|t}\right\} \oplus \mathcal{H}^N \subseteq \mathcal{X}_{\text{Glo}}^{\text{f}} \oplus \mathcal{H}^N \subseteq \mathcal{Y}$. It follows that

$$V_i^{\text{f}}\big(\bar{\boldsymbol{x}}_{i,t+N+1|t+1}\big) \leq \tilde{V}_i^{\text{f}}\left(\bar{\boldsymbol{x}}_{i,t+N|t+1}\right) \tag{B.80}$$

$$\leq \tilde{V}_i^{\text{f}}\left(\bar{\boldsymbol{x}}_{i,t+N|t}\right) + \tilde{\Delta}_i^{\text{f}} \tag{B.81}$$

$$= \alpha_{i,t+1} \tag{B.82}$$

where (6.7a) has been applied for the first and (6.20) for the second inequality. If the update rule for $\alpha_{i,t+1}$ is chosen to (6.21), then $\bar{\boldsymbol{x}}_{i,t+N+1|t+1} \in \mathcal{X}_i^{\mathrm{f}}(\alpha_{i,t+1})$. Moreover, it has to be ensured that $\prod_{i=1}^{M} \mathcal{X}_i^{\mathrm{f}}(\alpha_{i,t+1}) \subseteq \mathcal{X}_{\mathrm{Glo}}^{\mathrm{f}}$. For the global function,

$$V^{\mathrm{f}}\left(\bar{\boldsymbol{x}}_{t+N+1|t+1}\right) = \sum_{i=1}^{M} V_i^{\mathrm{f}}\left(\bar{\boldsymbol{x}}_{i,t+N+1|t+1}\right) \tag{B.83}$$

$$\leq \sum_{i=1}^{M} \alpha_{i,t+1} \tag{B.84}$$

holds. What is left to show is that $\sum_{i=1}^{M} \alpha_{i,t+1} \leq \alpha$. Summing $\alpha_{i,t+1}$ over all subsystems leads to

$$\sum_{i=1}^{M} \alpha_{i,t+1} = \sum_{i=1}^{M} \tilde{V}_i^{\mathrm{f}}\left(\bar{\boldsymbol{x}}_{i,t+N|t}\right) + \tilde{\Delta}_i^{\mathrm{f}}$$
$$\leq V^{\mathrm{f}}\left(\bar{\boldsymbol{x}}_{t+N|t}\right) - q\left(\bar{\boldsymbol{x}}_{t+N|t}\right) + \tilde{\Delta}^{\mathrm{f}}.$$

Due to Assumption 6.3.3, $\bar{\boldsymbol{x}}_{t+N|t} \in \mathcal{X}_{\mathrm{Glo}}^{\mathrm{f}}$ implies that $\sum_{i=1}^{M} \alpha_{i,t+1} \leq \alpha$. Moreover, note that $\alpha_{i,t} > 0$ for all $t \in \mathbb{N}$ because $\alpha_{i,t} \geq \tilde{\Delta}_i^{\mathrm{f}} > 0$. Thus, the sets $\mathcal{X}_i^{\mathrm{f}}(\alpha_i)$ are non-empty. Since $\sum_{i=1}^{M} \alpha_{i,0} \leq \alpha$, it follows by induction that $V^{\mathrm{f}}\left(\bar{\boldsymbol{x}}_{t+N+1|t+1}\right) \leq \alpha \; \forall t > 0$. It follows that $\boldsymbol{V}_{t+1}^{\mathrm{shift}}$ is a feasible input sequence. This completes the proof. □

Bibliography

[ABB11] A. Alessio, D. Barcelli, and A. Bemporad. Decentralized model predictive control of dynamically coupled linear systems. *Journal of Process Control*, 21(5):705–714, 2011.

[AGL15] S. Al-Areqi, D. Görges, and S. Liu. Event-based control and scheduling codesign: Stochastic and robust approaches. *IEEE Transactions on Automatic Control*, 60(5):1291–1303, 2015.

[Al-16] S. Al-Areqi. *Investigation on Robust Codsign Methods for Networked Control Systems*. PhD thesis, Fachbereich Elektrotechnik und Informationstechnik der Technischen Universitat Kaiserslautern, 2016.

[ANAS10] M. Arnold, R. R. Negenborn, G. Andersson, and B. De Schutter. *Intelligent Infrastructures*, chapter Distributed Predictive Control for Energy Hub Coordination in Coupled Electricity and Gas Networks, pages 235–273. Springer Netherlands, 2010.

[ApS17] MOSEK ApS. *The MOSEK optimization toolbox for MATLAB manual. Version 8.1.*, 2017.

[ÅW97] K. J. Åström and B. Wittenmark. *Computer-controlled systems: theory and design*. Prentice-Hall, 1997.

[BB12] D. Bernardini and A. Bemporad. Energy-aware robust model predictive control based on noisy wireless sensors. *Automatica*, 48:36–44, 2012.

[BBBL18] F. Berkel, J. Bleich, M. Bell, and S. Liu. A distributed voltage controller for medium voltage grids with storage-containing loads. In *Proceedings of the 44th Annual Conference of the IEEE Industrial Electronics Society (IECON2018)*, 2018.

[BBL17] M. Bell, F. Berkel, and S. Liu. Optimal distributed balancing control for three-phase four-wire low voltage grids. In *Proceedings of the IEEE International Conference on Smart Grid Communications (SGC 2017)*, Dresden/Germany, 2017.

[BBL19] M. Bell, F. Berkel, and S. Liu. Real-time distributed control of low voltage grids with dynamic optimal power dispatch of renewable energy sources. *IEEE Transactions on Sustainable Energy*, 10(1):417–425, 2019.

[BBM17] F. Borrelli, A. Bemporad, and M. Morari. *Predictive control for linear and hybrid systems.* Cambridge University Press, 2017.

[BC08] M. Bauer and I. K. Craig. Economic assessment of advanced process control–a survey and framework. *Journal of process control*, 18(1):2–18, 2008.

[BC18] H. S. Bidgoli and T. Van Cutsem. Combined local and centralized voltage control in active distribution networks. *IEEE Transactions on Power Systems*, 33(2):1374–1384, 2018.

[BCBL18] F. Berkel, S. Caba, J. Bleich, and S. Liu. A modeling and distributed MPC approach for water distribution networks. *Control Engineering Practice*, 81:199–206, 2018.

[BCCC11] S. Bolognani, G. Cavraro, F. Cerruti, and A. Costabeber. A linear dynamic model for microgrid voltages in presence of distributed generation. In *First International Workshop on Smart Grid Modeling and Simulation (SGMS)*, pages 31–36. IEEE, 2011.

[BFB+16] M. Bell, S. Fuchs, F. Berkel, S. Liu, and D. Görges. A privacy preserving negotiation-based control scheme for low voltage grids. In *Proceedings of the IEEE International Symposium on Industrial Electronics (ISIE 2016)*, Santa Clara/USA, 2016.

[BFS14] G. Betti, M. Farina, and R. Scattolini. Realization issues, tuning, and testing of a distributed predictivecontrol algorithm. *Journal of Process Control*, 24(4):424–434, 2014.

[BGB12] R. Bourdais, H. Guéguen, and A. Belmiloudi. Distributed model predictive control for a class of hybrid system based on lagrangian relaxation. *IFAC Proceedings Volumes*, 45(9):46–51, 2012.

[BGL13] F. Berkel, D. Görges, and S. Liu. Load-frequency control, economic dispatch and unit commitment in smart mircogrids based on hierarchical model predictive control. In *Proceedings of the 52nd IEEE Conference on Decision and Control (CDC 2013)*, pages 2326–2333, Florence/Italy, 2013.

[BHA17] F. D. Brunner, W. P. M. H. Heemels, and F. Allgöwer. Robust event-triggered MPC with guaranteed asymptotic bound and average sampling rate. *IEEE Transactions on Automatic Control*, 2017.

[BHT17] P. R. Baldivieso Monasterios, B. Hernandez, and P. A. Trodden. Nested distributed MPC. In *Proceedings of the 20th IFAC World Congress*, 2017.

[BL18a] F. Berkel and S. Liu. An event-triggered cooperation approach for robust distributed model predictive control. *IFAC Journal of Systems and Control*, 6:16–24, 2018.

[BL18b] F. Berkel and S. Liu. An event-triggered output-based model predictive control strategy. *Accepted for publication in IEEE Transactions on Control of Network Systems*, 2018.

[BL18c] F. Berkel and S. Liu. Non-iterative distributed model predictive control with event-triggered communication. In *Proceedings of the American Control Conference (ACC 2018)*, pages 2344–2349, Milwaukee, USA, 2018.

[BM99] A. Bemporad and M. Morari. Control of systems integrating logic, dynamics, and constraints. *Automatica*, 35(3):407–427, 1999.

[BMDP02] A. Bemporad, M. Morari, V. Dua, and E. N. Pistikopoulos. The explicit linear quadratic regulator for constrained systems. *Automatica*, 38(1):3–20, 2002.

[BPC+10] S. Boyd, N. Parikh, E. Chu, B. Peleato, and J. Eckstein. Distributed optimization and statistical learning via the alternating direction method of multipliers. *Foundations and Trends in Machine Learning*, 3:1–222, 2010.

[BT97] D. P. Bertsekas and J. N. Tsitsiklis. *Parallel and Distributed Computation: Numerical Methods*. Athena Scientific, 1997.

[BU94] M. A. Brdys and B. Ulanicki. *Operational control of water systems: structures, algorithms, and applications*. Prentice Hall, 1994.

[BWLG17] F. Berkel, B. Watkins, S. Liu, and D. Görges. Output-based event-triggered model predictive control for networked control systems. In *Proceedings of the 20th IFAC World Congress*, pages 9281–9286, Toulouse, France, 2017.

[Cas14] C. G. Cassandras. The event-driven paradigm for control, communication and optimization. *Journal of Control and Decision*, 1(1):3–17, 2014.

[CB13] E. F. Camacho and C. A. Bordons. *Model predictive control*. Springer Science & Business Media, 2013.

[CD09] E. Camponogara and L. B. De Oliveira. Distributed optimization for model predictive control of linear-dynamic networks. *IEEE Transactions on Systems, Man, and Cybernetics-Part A: Systems and Humans*, 39(6):1331–1338, 2009.

[CJKT02] E. Camponogara, D. Jia, B. H. Krogh, and S. Talukdar. Distributed model predictive control. *IEEE Control Systems*, 22(1):44–52, 2002.

[CJMZ16] C. Conte, C. N. Jones, M. Morari, and M. N. Zeilinger. Distributed synthesis and stability of cooperative distributed model predictive control for linear systems. *Automatica*, 69:117–125, 2016.

[CLL16] S. Caba, M. Lepper, and S. Liu. Nonlinear controller and observer design for centrifugal pumps. In *Proceedings of the IEEE Conference on Control Applications (CCA 2016)*, pages 569–574, Buenos Aires, Argentinia, 2016.

[CR80] C. R. Cutler and B. L. Ramaker. Dynamic matrix control - a computer control algorithm. In *Proceedings of the Joint Automatic Control Conference*, number 17, page 72, 1980.

[CRZ01] L. Chisci, J. A. Rossiter, and G. Zappa. Systems with persistent disturbances: predictive control with restricted constraints. *Automatica*, 37(7):1019–1028, 2001.

[CSML13] P. D. Christofides, R. Scattolini, D. Muñoz del la Peña, and J. Liu. Distributed model predictive control: A tutorial review and future research directions. *Computers and Chemical Engineering*, 51:21–41, 2013.

[CSZ+12] C. Conte, T. Summers, M. N. Zeilinger, M. Morari, and C. N. Jones. Computational aspects of distributed optimization in model predictive control. In *Proceedings of the 51st Annual Conference on Decision and Control (CDC)*, pages 6819–6824. IEEE, 2012.

[DH17] V. Dolk and W. P. M. H. Heemels. Event-triggered control systems under packet losses. *Automatica*, 80:143–155, 2017.

[DKD11] M. D. Doan, T. Keviczky, and B. De Schutter. A dual decomposition-based optimization method with guaranteed primal feasibility for hierarchical MPC problems. *IFAC Proceedings Volumes*, 44(1):392–397, 2011.

[DM06] W. B. Dunbar and R. M. Murray. Distributed receding horizon control for multi-vehicle formation stabilization. *Automatica*, 42(4):549–558, 2006.

[DS09] J. Dold and O. Stursberg. Distributed predictive control of communicating and platooning vehicles. In *Proceedings of the 48th IEEE Conference on Decision and Control, 2009 held jointly with the 2009 28th Chinese Control Conference (CDC/CCC 2009)*, pages 561–566. IEEE, 2009.

[Dun07] W. B. Dunbar. Distributed receding horizon control of dynamically coupled nonlinear systems. *IEEE Transactions on Automatic Control*, 50(7), 2007.

[EDK10] A. Eqtami, D. V. Dimarogonas, and K. J. Kyriakopoulos. Event-triggered control for discrete-time systems. In *Proceedings of the American Control Conference (ACC 2010)*, pages 4719–4724, Baltimore, MD, USA, 2010.

[EDK11a] A. Eqtami, D. V. Dimarogonas, and K. J. Kyriakopoulos. Event-triggered strategies for decentralized model predictive controllers. In *Proceedings of the 18th IFAC World Congress*, pages 10068–10073, Milano, Italy, 2011.

[EDK11b] A. Eqtami, D. V. Dimarogonas, and K. J. Kyriakopoulos. Novel event-triggered strategies for model predictive controllers. In *Proceedings of the 50th IEEE Conference on Decision and Control and European Control Conference (CDC/ECC 2011)*, pages 3392–3397, Orlando, USA, 2011.

[EDK12] A. Eqtami, D. V. Dimarogonas, and K. J. Kyriakopoulos. Event-based model predictive control for the cooperation of distributed agents. In *Proceedings of the American Control Conference (ACC 2012)*, pages 6473–6478, Montreal, 2012.

[Fai07] Y. R. Faizulkhakov. Time synchronization methods for wireless sensor networks: A survey. *Programming and Computer Software*, 33(4):214–226, 2007.

[FGM+15] M. Farina, A. Guagliardi, F. Mariani, C. Sandroni, and R. Scattolini. Model predictive control of voltage profiles in MV networks with distributed generation. *Control Engineering Practice*, 34:18–29, 2015.

[FMP+08] E. Franco, L. Magni, T. Parisini, M. M. Polycarpou, and D. M. Raimondo. Cooperative constrained control of distributed agents with nonlinear dynamics and delayed information exchange: A stabilizing receding-horizon approach. *IEEE Transactions on Automatic Control*, 53(1):324–338, 2008.

[FS12] M. Farina and R. Scattolini. Distributed predictive control: A non-cooperative algorithm with neighbor-to-neighbor communication for linear systems. *Automatica*, 48(6):1088–1096, 2012.

[FV09] R. Findeisen and P. Varutti. Stabilizing nonlinear predictive control over nondeterministic communication networks. In *Nonlinear model predictive control*, pages 167–179. Springer, 2009.

[GC10] R. A. Gupta and M.-Y. Chow. Networked control system: Overview and research trends. *IEEE transactions on industrial electronics*, 57(7):2527–2535, 2010.

[GOMP17] J. M. Grosso, C. Ocampo-Martínez, and V. Puig. A distributed predictive control approach for periodic flow-based networks: application to drinking water systems. *International Journal of Systems Science*, 48(14):3106–

3117, 2017.

[Gör17] D. Görges. Model predictive control. Lecture Notes University of Kaiserslautern, Chapter 1, Slide 22, October 2017.

[GP11] L. Grüne and J. Pannek. *Nonlinear model predictive control.* Springer, 2011.

[GR14] P. Giselsson and A. Rantzer. On feasibility, stability and performance in distributed model predictive control. *IEEE Transactions on Automatic Control*, 59(4):1031–1036, 2014.

[Gro15] J. M. Grosso. *On Model Predictive Control for Economic and Robust Operation of Generalised Flow-based Networks.* PhD thesis, Institut de Robotica i Informatica Industrial Universitat Politecnica de Catalunya. BarcelonaTech, 2015.

[GS13a] D. Gross and O. Stursberg. Distributed predictive control for a class of hybrid systems with event-based communication. *IFAC Proceedings Volumes*, 46(27):383–388, 2013.

[GS13b] D. Gross and O. Stursberg. On the convergence rate of a jacobi algorithm for cooperative distributed MPC. In *Proceedings of the 52nd IEEE Conference on Decision and Control (CDC 3013)*, pages 1508–1513. IEEE, 2013.

[GS16] D. Gross and O. Stursberg. A cooperative distributed MPC algorithm with event-basd communication and parallel optimization. *IEEE Transactions on Control of Network Systems*, 3(3):275–285, 2016.

[HAD14] K. Hashimoto, S. Adachi, and D. V. Dimarogonas. Distributed event-based model predictive control for multi-agent systems under disturbances. In *Proceedings of the 7th International Conference on NETwork Games, COntrol and OPtimization (NetGCoop 2014)*, pages 255–261. IEEE, 2014.

[HAD15] K. Hashimoto, S. Adachi, and D. V. Dimarogonas. Time-constrained event-triggered model predictive control for nonlinear continuous-time systems. In *Proceedings of the 54th IEEE Conference on Decision and Control (CDC 2015)*, pages 4326–4331. IEEE, 2015.

[HDT13] W. P. M. H. Heemels, M. C. F. Donkers, and A. R. Teel. Periodic event-triggered control for linear systems. *IEEE Transactions on Automatic Control*, 58(4):847–861, 2013.

[HJT12] W. P. M. H. Heemels, K. Johansson, and P. Tabuada. An introduction to event-triggered and self-triggered control. In *Proceedings of the 51st IEEE Conference on Decision and Control (CDC 2012)*, pages 3270–3285, Maui,

USA, 2012.

[HKJM13] M. Herceg, M. Kvasnica, C. N. Jones, and M. Morari. Multi-Parametric Toolbox 3.0. In *Proceedings of the European Control Conference (ECC 2013)*, pages 502–510, Zurich, Switzerland, 2013. URL: http://control.ee.ethz.ch/ mpt.

[HNX07] J. P. Hespanha, P. Naghshtabrizi, and Y. Xu. A survey of recent results in networked control systems. *Proceedings of the IEEE*, 95(1):138–162, 2007.

[HS15] N. He and D. Shi. Event-based robust sampled-data model predictive control: A non-monotonic lyapunov function approach. *IEEE Transactions on Circuits and Systems I: Regular Papers*, 62(10):2555–2564, 2015.

[HT16] B. Hernandez and P. Trodden. Distributed model predictive control using a chain of tubes. In *Proceedings of the UKACC 11th International Conference on Control*, Belfast, UK, 2016.

[HVG+10] W. P. M. H Heemels, N. Van de Wouw, R. H. Gielen, M. C. F. Donkers, L. Hetel, S. Olaru, M. Lazar, J. Daafouz, and S. Niculescu. Comparison of overapproximation methods for stability analysis of networked control systems. In *Proceedings of the 13th ACM international conference on Hybrid systems: computation and control*, pages 181–190. ACM, 2010.

[IBM14] IBM. *IBM ILOG Optimization Studio 12.6*, 2014.

[IFM17] G. P. Incremona, A. Ferrara, and L. Magni. Asynchronous networked MPC with ISM for uncertain nonlinear systems. *IEEE Transactions on Automatic Control*, 62(9):4305–4317, 2017.

[JDM15] M. Jost, M. Schulze Darup, and M. Mönnigmann. Optimal and suboptimal event-triggering in linear model predictive control. In *Proceedings of the European Control Conference (ECC 2015)*, pages 1153–1158, Linz, Austria, 2015.

[JTN05] K. H. Johansson, M. Törngren, and L. Nielsen. Vehicle applications of controller area network. In *Handbook of networked and embedded control systems*, pages 741–765. Springer, 2005.

[KBM96] M. V. Kothare, V. Balakrishnan, and M. Morari. Robust constrained model predictive control using linear matrix inequalities. *Automatica*, 32(10):1361–1379, 1996.

[KF12] M. Kögel and R. Findeisen. Cooperative distributed MPC using the alternating direction multiplier method. *IFAC Proceedings Volumes*, 45(15):445–450, 2012.

[KF16] M. Kögel and R. Findeisen. Output feedback MPC with send-on-delta measurements for uncertain systems. *IFAC-PapersOnLine*, 49(22):145–150, 2016.

[KG88] S.S. Keerthi and E.G. Gilbert. Optimal infinite-horizon feedback laws for a general class of constrained discrete-time systems: Stability and moving-horizon approximations. *Journal of Optimization Theory and Applications*, 57(2):265–293, 1988.

[KG98] I. Kolmanovsky and E.G. Gilbert. Theory and computation of disturbance invariant sets for discrete-time linear systems. *Mathematical Problems in Engineering*, 4(4):317–367, 1998.

[KM04] E. C. Kerrigan and J. M. Maciejowski. Feedback min-max model predictive control using a single linear program: robust stability and the explicit solution. *International Journal of Robust and Nonlinear Control: IFAC-Affiliated Journal*, 14(4):395–413, 2004.

[LA14] J. Lee and D. Angeli. Cooperative economic model predictive control for linear systems with convex objectives. *European Journal of Control*, 20(3):141–151, 2014.

[LAC02] D. Limon, T. Alamo, and E. Camacho. Input-to-state stable MPC for constrained discrete-time nonlinear systems with bounded additive uncertainties. In *Proceedings of the 41st IEEE Conference on Decision and Control (CDC 2002)*, pages 4619–4624, Las Vegas, USA, 2002.

[LASC06] D. Limon, T. Alamo, F. Salas, and E. F. Camacho. Input to state stability of min–max mpc controllers for nonlinear systems with bounded uncertainties. *Automatica*, 42(5):797–803, 2006.

[LHJ13] D. Lehmann, E. Henriksson, and K. H. Johansson. Event-triggered model predictive control of discrete-time linear systems subject to disturbances. In *Proceedings of the European Control Conference (ECC 2013)*, pages 1156–1161, Zurich, Switzerland, 2013.

[LM67] E. B. Lee and L. Markus. Foundations of optimal control theory. Technical report, Mineapolis Center for Control Science, Minnesota University, 1967.

[LMC09] J. Liu, D. Muñoz de la Peña, and P. D. Christofides. Distributed model predictive control of nonlinear process systems. *AIChE Journal*, 55(5):1171–1184, 2009.

[LMC10] J. Liu, D. Muñozde la Peña, and P. D. Christofides. Distributed model predictive control of nonlinear systems subject to asynchronous and delayed measurements. *Automatica*, 46(1):52–61, 2010.

[LMHA08] M. Lazar, D. Muñoz De La Peña, W. P. M. H. Heemels, and T. Alamo. On input-to-state stability of min–max nonlinear model predictive control. *Systems & Control Letters*, 57(1):39–48, 2008.

[Löf04] J. Löfberg. YALMIP: A toolbox for modeling and optimization in MATLAB. In *Proceedings of the CACSD Conference*, pages 284–289, Taipei, Taiwan, 2004. URL: http://users.isy.liu.se/johanl/yalmip.

[LPM+14] D. Limon, M. Pereira, D. Muñoz De La Peña, T. Alamo, and J. M. Grosso. Single-layer economic model predictive control for periodic operation. *Journal of Process Control*, 24(8):1207–1224, 2014.

[LS13a] H. Li and Y. Shi. Distributed model predictive control of constrained nonlinear systems with communication delays. *Systems & Control Letters*, 62(10):819–826, 2013.

[LS13b] H. Li and Y. Shi. Output feedback predictive control for constrained linear systems with intermittent measurements. *Systems & Control Letters*, 62(4):345–354, 2013.

[LS14] H. Li and Y. Shi. Event-triggered robust model predictive control of continuous-time nonlinear systems. *Automatica*, 50:1507–1513, 2014.

[LYSW15] H. Li, W. Yan, Y. Shi, and Y. Wang. Periodic event-triggering in distributed receding horizon control of nonlinear systems. *Systems & Control Letters*, 86:16–23, 2015.

[LZL+13] S. Liu, J. Zhang, J. Liu, Y. Feng, and G. Rong. Distributed model predictive control with asynchronous controller evaluations. *The Canadian Journal of Chemical Engineering*, 91(10):1609–1620, 2013.

[LZN17] L. Lu, Y. Zou, and Y. Niu. Event-driven robust output feedback control for constrained linear systems via model predictive control method. *Circuits, Systems, and Signal Processing*, 36(2):543–558, 2017.

[LZND10] S. Leirens, C. Zamora, R. R. Negenborn, and B. De Schutter. Coordination in urban water supply networks using distributed model predictive control. In *Proceedings of the American Control Conference (ACC 2010)*, pages 3957–3962, Baltimore, Maryland, 2010.

[LZNS10] S. Leirens, C. Zamora, R. R. Negenborn, and B. De Schutter. Coordination in urban water supply networks using distributed model predictive control. In *Proceedings of the American Control Conference (ACC)*, pages 3957–3962. IEEE, 2010.

[MA17] M. A. Müller and F. Allgöwer. Economic and distributed model predictive control: recent developments in optimization-based control. *SICE Journal of Control, Measurement, and System Integration*, 10(2):39–52, 2017.

[Mac02] J. M. Maciejowski. *Predictive control: with constraints*. Pearson education, 2002.

[May14] D. Q. Mayne. Model predictive control: Recent developments and future promise. *Automatica*, 50(12):2967–2986, 2014.

[MBA05] D. Muñoz De La Peña, A. Bemporad, and T. Alamo. Stochastic programming applied to model predictive control. In *Proceedings of the 44th IEEE Conference on Decision and Control and European Control Conference (CDC/ECC)*, pages 1361–1366. IEEE, 2005.

[MBDB10] P.-D. Moroşan, R. Bourdais, D. Dumur, and J. Buisson. Distributed model predictive control for building temperature regulation. In *Proceedings of the American Control Conference (ACC 2010)*, pages 3174–3179. IEEE, 2010.

[MC14] M. Mazo Jr. and M. Cao. Asynchronous decentralized event-triggered control. *Automatica*, 50(12):3197–3203, 2014.

[MDK12] S. Maniatopoulos, D. V. Dimarogonas, and K. J. Kyriakopoulos. A decentralized event-based predictive navigation scheme for air-traffic control. In *Proceedings of the American Control Conference (ACC 2012)*, pages 2503–2508. IEEE, 2012.

[Mey00] C. D. Meyer. *Matrix analysis and applied linear algebra*, volume 71. Siam, 2000.

[ML18] X. Mi and S. Li. Event-triggered mpc design for distributed systems with network communications. *IEEE/CAA Journal of Automatica Sinica*, 5(1):240–250, 2018.

[MM15] M. S. Mahmoud and A. M. Memon. Aperiodic triggering mechanisms for networked control systems. *Information Sciences*, 296:282–306, 2015.

[MMC09] J. M. Maestre, D. Muñoz De La Peña, and E. F. Camacho. Distributed MPC: a supply chain case study. In *Proceedings of the 48th IEEE Conference on Decision and Control held jointly with the 28th Chinese Control Conference (CDC/CCC 2009)*, pages 7099–7104. IEEE, 2009.

[MMCA11] J. M. Maestre, D. Muñoz De La Peña, E. F. Camacho, and T. Alamo. Distributed model predictive control based on agent negotiation. *Journal of Process Control*, 21(5):685–697, 2011.

[MN+14] J. M. Maestre, R. R. Negenborn, et al. *Distributed model predictive control made easy*, volume 69. Springer, 2014.

[MRRS00] D. Q. Mayne, J. B. Rawlings, C. V. Rao, and P. O. M. Scokaert. Constrained model predictive control: Stability and optimality. *Automatica*, 36(6):789–814, 2000.

[MSR05] D. Q. Mayne, M. M. Seron, and S. V. Rakovic. Robust model predictive control of constrained linear systems with bounded disturbances. *Automatica*, 41(2):219–224, 2005.

[MT07] J. R. Moyne and D. M. Tilbury. The emergence of industrial control networks for manufacturing control, diagnostics, and safety data. *Proceedings of the IEEE*, 95(1):29–47, 2007.

[MUA14] M. C. Meinel, M. Ulbrich, and S. Albrecht. A class of distributed optimization methods with event-triggered communication. *Computational Optimization and Applications*, 57:517–553, 2014.

[NC13] I. Necoara and D. Clipici. Efficient parallel coordinate descent algorithm for convex optimization problems with separable constraints: application to distributed MPC. *Journal of Process Control*, 23(3):243–253, 2013.

[NM14] R. R. Negenborn and J. M. Maestre. Distributed model predictive control - an overview and roadmap of future research opportunities. *IEEE Control Systems Magazine*, 34(4):87–97, 2014.

[NvKD09] R. R. Negenborn, P.-J. van Overloop, T. Keviczky, and B. De Schutter. Distributed model predictive control of irrigation canals. *Networks and Heterogeneous Media*, 4(2), 2009.

[NVKP14] Q. T. Nguyen, V. Veselý, A. Kozáková, and P. Pakshin. Networked robust predictive control systems design with packet loss. *Journal of Electrical Engineering*, 65(1):3–11, 2014.

[OBPB12] C. Ocampo-Martinez, D. Barcelli, V. Puig, and A. Bemporad. Hierarchical and decentralised model predictive control of drinking water networks: Application to barcelona case study. *IET Control Theory & Applications*, 6(1):62–71, 2012.

[PP11] G. Pin and T. Parisini. Networked predictive control of uncertain constrained nonlinear systems: Recursive feasibility and input-to-state stability analysis. *IEEE Transactions on Automatic Control*, 56(1):72–87, 2011.

[Pro63] A. I. Propoı. Application of linear programming methods for the synthesis of automatic sampled-data systems. *Avtomat. i Telemeh*, 24:912–920, 1963.

[QB03] S. J. Qin and T. A. Badgwell. A survey of industrial model predictive control technology. *Control Engineering Practice*, 11(7):733–764, 2003.

[QMFC15] D. E. Quevedo, P. K. Mishra, R. Findeisen, and D. Chatterjee. A stochastic model predictive controller for systems with unreliable communications. *IFAC-PapersOnLine*, 48(23):57–64, 2015.

[RA09] J. B. Rawlings and R. Amrit. Optimizing process economic performance using model predictive control. In *Nonlinear model predictive control*, pages 119–138. Springer, 2009.

[RAB12] J. B. Rawlings, D. Angeli, and C. N. Bates. Fundamentals of economic model predictive control. In *Proceedings of the 51st Conference on Decision and Control (CDC 2012)*, pages 3851–3861, Maui, USA, 2012.

[RBFT13] S. Riverso, A. Battocchio, and G. Ferrari-Trecate. *PnPMPC toolbox*, 2013.

[RFFT13] S. Riverso, M. Farina, and G. Ferrari-Trecate. Plug-and-play decentralized model predictive control for linear systems. *IEEE Transactions on Automatic Control*, 58(10):2608–2614, 2013.

[RFT12] S. Riverso and G. Ferrari-Trecate. Tube-based distributed control of linear constrained systems. *Automatica*, 48(11):2860–2865, 2012.

[RG17] R. Rostami and D. Görges. Distributed model predictive control with event-based optimization. *IFAC-PapersOnLine*, 50(1):8933–8938, 2017.

[RH07] A. Richards and J. P. How. Robust distributed model predictive control. *International Journal of control*, 80(9):1517–1531, 2007.

[RM09] J. B. Rawlings and D. Q. Mayne. *Model Predictive Control - Theory and Design*. Nob Hill Publishing, 2009.

[RMS07] D. M. Raimondo, L. Magni, and R. Scattolini. Decentralized MPC of non-linear systems: an input-to-state stability approach. *International Journal of Robust and Nonlinear Control*, 17:1651–1667, 2007.

[RRTP78] J. Richalet, A. Rault, J. L. Testud, and J. Papon. Model predictive heuristic control. *Automatica*, 14(5):413–428, 1978.

[SBK05] B. Sundararaman, U. Buy, and A. D. Kshemkalyani. Clock synchronization for wireless sensor networks: a survey. *Ad hoc networks*, 3(3):281–323, 2005.

[SC07] R. Scattolini and P. Colaneri. Hierarchical model predictive control. In *Proceedings of the 46th IEEE Conference on Decision and Control*, pages 4803–4808. IEEE, 2007.

[Sca09] R. Scattolini. Architectures for distributed and hierarchical model predictive control - a review. *Journal of Process Control*, 19(5):723–731, 2009.

[SCBM15] V. Spudić, C. Conte, M. Baotić, and M. Morari. Cooperative distributed model predictive control for wind farms. *Optimal Control Applications and Methods*, 36(3):333–352, 2015.

[SF11] L. Schenato and F. Fiorentin. Average timesynch: A consensus-based protocol for clock synchronization in wireless sensor networks. *Automatica*, 47(9):1878–1886, 2011.

[Šil91] D. Šiljak. *Decentralized Control of Complex Systems.* Academic Press, Cambridge, MA, 1991.

[SL12] T. H. Summers and J. Lygeros. Distributed model predictive consensus via the alternating direction method of multipliers. In *Proceedings of the 50th Annual Allerton Conference on Communication, Control, and Computing*, pages 79–84. IEEE, 2012.

[SLH10] J. Sijs, M. Lazar, and W. P. M. H. Heemels. On integration of event-based estimation and robust MPC in a feedback loop. In *Proceedings of the 13th International Conference on Hybrid Systems*, pages 31–40, Stockholm, Sweden, 2010.

[SSF+07] L. Schenato, B. Sinopoli, M. Franceschetti, K. Poolla, and S. S. Sastry. Foundations of control and estimation over lossy networks. *Proceedings of the IEEE, Special Issue on Networked Control Systems*, 95(1):163–187, 2007.

[SVR+10] B. T. Stewart, A. N. Venkat, J. B. Rawlings, S. J. Wright, and G. Pannocchia. Cooperative distributed model predictive control. *System & Control Letters*, 59(8):460–469, 2010.

[SWR11] B. T. Stewart, S. L. Wright, and J. B. Rawlings. Cooperative distributed model predictive control for nonlinear systems. *Journal of Process Control*, 21(5):698–704, 2011.

[TM17] P. A. Trodden and J. M. Maestre. Distributed predictive control with minimization of mutual disturbances. *Automatica*, 77:31–43, 2017.

[TR10] P. A. Trodden and A. Richards. Distributed model predictive control of linear systems with persistent disturbances. *International Journal of Control*, 83(8):1653–1663, 2010.

[TR13] P. A. Trodden and A. Richards. Cooperative distributed MPC of linear systems with coupled constraints. *Automatica*, 49(2):479–487, 2013.

[VF11] P. Varutti and R. Findeisen. Event-based NMPC for networked control systems over UDP-like communication channels. In *Proceedings of the American Control Conference (ACC 2011)*, pages 3166–3171. IEEE, 2011.

[VFK+10] P. Varutti, T. Faulwasser, B. Kern, M. Kögel, and R. Findeisen. Event-based reduced-attention predictive control for nonlinear uncertain systems. In *Proceedings of the IEEE Multi-Conference on Systems and Control (MSC 2010)*, pages 1085–1090, Yokohama, Japan, 2010.

[VHRW08] A. N. Venkat, I. A. Hiskens, J. B. Rawlings, and S. J. Wright. Distributed MPC strategies with application to power system automatic generation control. *IEEE Transactions on Control Systems Technology*, 16(6):1192–1206, 2008.

[WBAL16a] B. Watkins, F. Berkel, S. Al-Areqi, and S. Liu. Distributed control of constrained linear systems using a resource aware communication strategy for networks with MAC. In *Proceedings of the IEEE Multi-Conference on Systems and Control (MSC 2016)*, Buenos Aires/Argentina, 2016.

[WBAL16b] B. Watkins, F. Berkel, S. Al-Areqi, and S. Liu. Event-based distributed control of dynamically coupled and constrained linear systems. In *Proceedings of the 55th IEEE Conference on Decision and Control (CDC 2016)*, pages 6074–6079, Las Vegas/USA, 2016.

[WO16] Z. Wang and C. J. Ong. Distributed MPC of constrained linear systems with online decoupling of the terminal constraint. *System & Control Letters*, 88:14–23, 2016.

[Zam08] S. Zampieri. Trends in networked control systems. *IFAC Proceedings Volumes*, 41(2):2886–2894, 2008.

[ZGD13] M. Zanon, S. Gros, and M. Diehl. A lyapunov function for periodic economic optimizing model predictive control. In *Proceedings of the 52nd IEEE Conference onDecision and Control (CDC 2013)*, pages 5107–5112, Florence, Italy, 2013.

[ZLL14] J. Zhang, S. Liu, and J. Liu. Economic model predictive control with triggered evaluations: State and output feedback. *Journal of Process Control*, 24(8):1197–1206, 2014.

Zusammenfassung

Beiträge zu ereignisbasierter und verteilter modellprädiktiver Regelung

Die Arbeit behandelt Ansätze zu ereignisbasierten und verteilten modellprädiktiven Regelungen von vernetzten und verteilten Regelungssystemen.

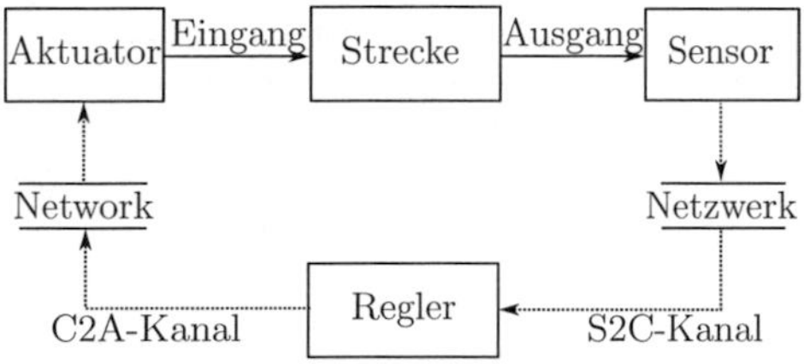

Abbildung B.1: Die Abbildung zeigt einen einzelnen Regelkreis, der über ein Netzwerk geschlossen ist. Der Sensor sendet den gemessenen Ausgang über den S2C-Kanal des Netzwerks. Der Regler schickt den Eingang über den C2A-Kanal an den Aktuator, der den Eingang auf das System gibt.

Ein Beispiel eines vernetztes Regelungssystems ist in Abbildung B.1 zu sehen. Sensor, Regler und Aktuator sind örtlich verteilt und tauschen Daten über ein Netzwerk miteinander aus. Die Nutzung des Netzwerks als Kommunikationsmedium bringt viele Vorteile gegenüber einer klassischen Festverdrahtung. So zeichnen sich Netzwerke beispielsweise durch geringere Installations- und Betriebskosten aus. Auf der anderen Seite treten durch das Netzwerk verursachte Verzögerungen und Paketverluste auf, welche im Regelungsdesign berücksichtigt werden müssen, damit die Stabilität des geschlossenen Regelkreises sichergestellt werden kann. Zudem werden die Netzwerke von mehreren Teilnehmern gleichzeitig benutzt und haben oft nur begrenzt verfügbare Ressourcen, weshalb eine sparsame Nutzung des Kommunikationsmediums durch die Regelung wünschenswert ist. Eine Regelungsmethode, die den Kommunikationsaufwand explizit berücksichtigt, ist die ereignisbasierte Regelung. Hier werden Daten nur bei einem Ereignis versendet.

Ein Ereignis wird in der Regel von einem sogenannten Ereignisgenerator erzeugt, welcher so entworfen wird, dass er dann Ereignisse generiert, wenn es aus Stabilitäts- oder Performanzgründen notwendig ist. Dies unterscheidet sich von der klassischen zeitbasierten Regelung, bei der an fest vordefinierten Zeitpunkten neue Informationen gesendet werden.

In einem verteilten Regelungssystem besteht das zu regelnde System aus vielen gekoppelten Teilsystemen. Aufgrund der hohen Anzahl und der örtlichen Verteilung der Teilsysteme ist es oft nicht möglich, das System zentral zu regeln, weshalb die Teilsysteme von einem lokalen Regler geregelt werden. Dabei tauschen die Regler Informationen untereinander aus, um die Gesamtperformanz des Systems zu erhöhen. In den letzten zwei Jahrzehnten wurde eine Vielzahl an verschiedenen Ansätzen für die Regelung von verteilten Regelungssystemen eingeführt. Sie unterscheiden sich nach Art der Kopplung der Teilsysteme, der vorhandenen Kommunikationsstruktur und der Absichten der einzelnen Systeme, Informationen untereinander auszutauschen.

Bezogen auf die ereignisbasierte modellprädiktive Reglung von vernetzten Regelungssystemen wurden in der Literatur bisher hauptsächlich Regelungen mit Zustandsrückführung betrachtet. In der Praxis werden jedoch meistens nicht der volle Zustand sondern nur gewisse Ausgänge gemessen. Zudem wurde die Behandlung von Netzwerkeffekten in ausgangsbasierten Verfahren bisher nicht betrachtet. Mit dieser Arbeit soll dazu beigetragen werden, diese vorhandenen Lücken zu schließen.

Bei der Betrachtung von verteilten modellprädiktiven Regelungen wurde bisher nur wenig auf den Kommunikationsaufwand zwischen den Teilsystemen geachtet. Insbesondere die Idee der ereignisbasierten Kommunikation zur Reduzierung der Kommunikation zwischen den Teilsystemen wurde in der Literatur bisher nur selten und nur für bestimmte Verfahren angewandt. Ein weiteres Ziel der Arbeit ist es diese vorhandene Lücke zu schließen, indem die Idee der ereignisbasierten Kommunikation auf weitere Ansätze übertragen werden soll. Dabei liegen vor allem Verfahren für Teilsysteme mit dynamischer Kopplung im Fokus.

Die genauen Inhalte der Kapitel der Arbeit sind wie folgt:

Kapitel 2 fasst die für die Arbeit notwendigen mathematischen Grundlagen und Konzepte der System- und Regelungstheorie zusammen. Weiterhin werden benötigte MPC-Konzepte vorgestellt.

Kapitel 3 stellt die beiden in der Arbeit betrachteten Szenarien, die dezentrale Regelung eines vernetzten Systems und die Regelung eines verteilten Systems vor. Zudem werden die betrachteten Systemklassen der zeitdiskreten linearen zeit-invarianten und periodisch zeitvarianten Systeme diskutiert. Die Modellierung des Kommunikationsnetzes und eine suboptimale robuste MPC Methode werden eingeführt.

In Kapitel 4 wird eine ausgangsbasierte dezentrale Regelung eines vernetzten zeitdiskreten Regelungssystems mit linearer zeitinvarianter Dynamik, das Beschränkungen unterliegt, betrachtet. Die Sensoren kommunizieren über ein Netzwerk mit den Reglern, welche wiederum über das Netzwerk mit den Aktuatoren kommunizieren.

Zunächst wird eine Regelung mit ereignisbasierter Kommunikation über den Sensor-Regler-Kanal des Netzwerks betrachtet. Dazu wird ein Ereignisgenerator im Sensor eingeführt, welcher den Ausgang der Strecke überwacht. Ein neuer Wert wird über das Netzwerk geschickt, wenn die Differenz zwischen aktuellem und zuletzt gesendetem Wert einen relativen oder absoluten Schwellwert überschreitet. Ein Beobachter rekonstruiert den Systemzustand mit Hilfe der empfangenen Daten. Der verwendete modellprädiktive Regler basiert auf robusten Verfahren.

Im nächsten Schritt wird zusätzlich eine ereignisbasierte Kommunikation zwischen den Reglern und Aktuatoren betrachtet. Dazu wird erneut ein Ereignisgenerator mit relativen und absoluten Schwellwerten eingesetzt. Wird ein Ereignis generiert, wird eine neue Eingangssequenz berechnet, welche dann über das Netzwerk zum Aktuator gesendet wird. Dort befindet sich ein Puffer, der auch Eingangsinformationen auf das System geben kann, wenn keine neue Information geschickt wird.

Für beide betrachteten Szenarien werden hinreichende Kriterien entwickelt, mit denen die rekursive Lösbarkeit, das Einhalten von Beschränkungen und die Stabilität des Ursprungs garantiert werden können. Des Weiteren wird im Zuge des Kapitels die Regelung von Abtastsystemen betrachtet und es wird diskutiert, wie die entworfene Regelung für den Fall der zeitveränderlichen Abtastung und Kommunikationsverzögerungen angepasst werden kann. Die entworfenen Verfahren werden in Simulationen für ein zentrales und ein dezentrales Regelungsszenario getestet.

Kapitel 5 betrachtet die verteilte Regelung eines zeitdiskreten Regelungssystems mit linearer zeitinvarianter Dynamik, das Beschränkungen unterliegt. Die entworfene Regelung arbeitet nicht iterativ, d.h. die Regler der Subsysteme kommunizieren maximal einmal im Abtastschritt miteinander.

Um den Kommunikationsaufwand zwischen den Teilsystemen zu reduzieren, werden in jedem Teilsystem Ereignisgeneratoren und Prädiktoren für die eigene und die Dynamik der Nachbarsysteme installiert. Diese Ereignisgeneratoren betrachten die Differenz des tatsächlichen lokalen Zustands zu dem des in den Nachbarsystemen prädizierten Zustands. Überschreitet die Differenz einen Schwellwert, welcher aus einem absoluten und einem relativen Anteil besteht, wird ein Ereignis ausgelöst und der aktuelle Zustand an die Nachbarsysteme geschickt. In diesem Zusammenhang werden Methoden der robusten modellprädiktiven Regelung verwendet.

Hinreichende Kriterien zur Sicherstellung der Konvergenz des Zustands zum Ursprung und dem Einhalten der Beschränkungen werden angegeben. Zudem wird ein Verfahren entworfen, mit dem die Regelung initialisiert werden kann.

Kapitel 6 betrachtet die verteilte Regelung eines zeitdiskreten Regelungssystems mit linearer zeitinvarianter Dynamik, das Beschränkungen unterliegt und einer Störung ausgesetzt ist.

Das vorgestellte Verfahren basiert auf Regelungsmethoden, die das Optimierungsproblem online mit verteilten iterativen Algorithmen lösen. Dabei arbeiten die Teilsysteme kooperativ zusammen. Um den Kommunikationsaufwand zu reduzieren, werden Ereignisgeneratoren entworfen, die die Kooperation nur dann auslösen, wenn die zuletzt kommunizierte und die aktuell prädizierte Trajektorie voneinander abweichen. Dadurch wird die Kooperation adaptiv zu den aktuellen Störungen im System angepasst und die Kommunikation wird indirekt verringert, weil der größte Kommunikationsaufwand beim kooperativen Lösen des Optimierungsproblems besteht. Um die rekursive Lösbarkeit sicherzustellen, wird ein robuster modellprädiktiver Regler mit einer zeitvarianten Endzustandsbeschränkung verwendet. Es werden eine Anpassungsregel für die Endzustandsbeschränkungen und hinreichende Kriterien eingeführt, die Stabilität des Ursprungs und die Einhaltung der Beschränkungen garantieren.

Kapitel 7 betrachtet zwei Systemklassen, die aus realen Anwendungen abgeleitet wurden. Das erste betrachtete Szenario zeigt die Regelung des Wassernetzes eines lokalen Betreibers in Kaiserslautern. Basierend auf einem nichtlinearen Modell wird ein vereinfachtes, für die Regelung verwendbares, Modell hergeleitet. Unter gewissen Voraussetzungen kann das System als periodisch zeitvariantes System aufgefasst werden. Zudem weist es eine kaskadierte Struktur mit Kopplung der Teilsysteme in der Dynamik über die Eingänge auf. Das Regelungsziel ist es, eine ökonomische Kostenfunktion zu minimieren, die hauptsächlich die Betriebskosten abbildet und nicht positiv definit bezüglich eines Referenzwertes ist. Durch die vorhandenen Pumpen mit diskreten Schaltstufen besitzt das vereinfachte Modell nicht-konvexe Beschränkungen.

Die zweite betrachtete Anwendung ist ein Szenario einer Spannungsregelung in einem Mittelspannungsnetz eines lokalen Betreibers in Kaiserslautern. Ein vereinfachtes lineares Modell wird für die Regelung hergeleitet. Unter gewissen Voraussetzungen kann das Modell als periodisch zeitvariantes Modell aufgefasst werden. Da ein Verbraucher ein Schaltverhalten aufweist, besitzt das Regelungsmodell nicht-konvexe Beschränkungen. Das Regelungsziel ist es, die Spannungen in den gewünschten Beschränkungen zu halten. Die Dynamiken der einzelnen Teilsysteme sind entkoppelt. Die Kopplung kommt durch die gemeinsamen Beschränkungen in der Spannung zustande.

Ein für beide Szenarien geeigneter Algorithmus wird hergeleitet. Er ist zugeschnitten auf die Charakteristiken der Anwendungen, insbesondere auf das periodisch zeitvariante Verhalten, die nicht konvexen Beschränkungen und die ökonomischen, nicht notwendigerweise positiv definiten Kostenfunktionen. Die Idee der Ereignisgeneratoren, die die

Abnahme der Kosten als Schwellwert verwenden, wird integriert, um den Kommunikationsaufwand zu reduzieren.

Kapitel 8 fasst die Arbeit zusammen und gibt einen Überblick über mögliche zukünftige Forschungsrichtungen.

Der Anhang beinhaltet ergänzendes Material und Informationen. Außerdem wurden längere Beweise aufgrund der besseren Übersichtlichkeit dorthin verlagert.

Curriculum Vitae

Personal Data

Name	Felix Berkel
E-Mail	berkel@eit.uni-kl.de

Education

1998 - 2007	Paul-von-Denis-Gymnasium Schifferstadt Degree: Abitur
2008 - 2012	University of Kaiserslautern Subject: Electrical Engineering Specialization: Automation Degree: Dipl.-Ing. (equivalent to M.Sc.) Thesis: "Hierarchische modellprädiktive Regelung von Smart Microgrids mit Elektrofahrzeugen"

Professional Experience

2013 - 2018	University of Kaiserslautern Department of Electrical and Computer Engineering Institute of Control Systems Research Associate

May 2019

In der Reihe *„Forschungsberichte aus dem Lehrstuhl für Regelungssysteme“*, herausgegeben von Steven Liu, sind bisher erschienen:

1	Daniel Zirkel	Flachheitsbasierter Entwurf von Mehrgrößenregelungen am Beispiel eines Brennstoffzellensystems ISBN 978-3-8325-2549-1, 2010, 159 S. 35.00 €
2	Martin Pieschel	Frequenzselektive Aktivfilterung von Stromoberschwingungen mit einer erweiterten modellbasierten Prädiktivregelung ISBN 978-3-8325-2765-5, 2010, 160 S. 35.00 €
3	Philipp Münch	Konzeption und Entwurf integrierter Regelungen für Modulare Multilevel Umrichter ISBN 978-3-8325-2903-1, 2011, 183 S. 44.00 €
4	Jens Kroneis	Model-based trajectory tracking control of a planar parallel robot with redundancies ISBN 978-3-8325-2919-2, 2011, 279 S. 39.50 €
5	Daniel Görges	Optimal Control of Switched Systems with Application to Networked Embedded Control Systems ISBN 978-3-8325-3096-9, 2012, 201 S. 36.50 €
6	Christoph Prothmann	Ein Beitrag zur Schädigungsmodellierung von Komponenten im Nutzfahrzeug zur proaktiven Wartung ISBN 978-3-8325-3212-3, 2012, 118 S. 33.50 €
7	Guido Flohr	A contribution to model-based fault diagnosis of electro-pneumatic shift actuators in commercial vehicles ISBN 978-3-8325-3338-0, 2013, 139 S. 34.00 €

8	Jianfei Wang	Thermal Modeling and Management of Multi-Core Processors	
		ISBN 978-3-8325-3699-2, 2014, 144 S.	35.50 €
9	Stefan Simon	Objektorientierte Methoden zum automatisierten Entwurf von modellbasierten Diagnosesystemen	
		ISBN 978-3-8325-3940-5, 2015, 197 S.	36.50 €
10	Sven Reimann	Output-Based Control and Scheduling of Resource-Constrained Processes	
		ISBN 978-3-8325-3980-1, 2015, 145 S.	34.50 €
11	Tim Nagel	Diagnoseorientierte Modellierung und Analyse örtlich verteilter Systeme am Beispiel des pneumatischen Leitungssystems in Nutzfahrzeugen	
		ISBN 978-3-8325-4157-6, 2015, 306 S.	49.50 €
12	Sanad Al-Areqi	Investigation on Robust Codesign Methods for Networked Control Systems	
		ISBN 978-3-8325-4170-5, 2015, 180 S.	36.00 €
13	Fabian Kennel	Beitrag zu iterativ lernenden modellprädiktiven Regelungen	
		ISBN 978-3-8325-4462-1, 2017, 180 S.	37.50 €
14	Yun Wan	A Contribution to Modeling and Control of Modular Multilevel Cascaded Converter (MMCC)	
		ISBN 978-3-8325-4690-8, 2018, 209 S.	37.00 €
15	Felix Berkel	Contributions to Event-triggered and Distributed Model Predictive Control	
		ISBN 978-3-8325-4935-0, 2019, 185 S.	36.00 €